RED

RED

MY UNCENSORED LIFE IN ROCK

SAMMY HAGAR

WITH JOEL SELVIN

FOREWORD BY
MICHAEL ANTHONY

*it*books

AN IMPRINT OF HARPERCOLLINS *PUBLISHERS*

FIRST EDITION

Designed by Renato Stanisic

Library of Congress Cataloging-in-Publication Data has been applied for.

ISBN 978-0-06-200928-9

11 12 13 14 15 OV/RRD 10 9 8

TO GLADYS

CONTENTS

CONTENTS

FOREWORD

BY MICHAEL ANTHONY

＊＊＊＊

I first saw the Van Halen brothers play when I was going to Arcadia High School, on the eastern edge of Los Angeles. It was during a student fair held on the football field, and the band was called Mammoth. It was just Eddie Van Halen on guitar, his brother Alex Van Halen on drums, and a guy named Mark Stone playing bass. Eddie did all the vocals. They played Cream, Grand Funk Railroad, and the Who. Eddie nailed every single note of every song, exactly like the records.

After high school, I belonged to a band called Snake. Pretty original name, I know. We opened a show for Van Halen at Pasadena High School. Now they had a vocalist. I remember sitting around the parking lot after the gig, talking with Eddie.

Fast forward to my second year at Pasadena City College, and through a mutual friend I got hooked up again with the Van Halen brothers. They wanted to get rid of their bass player and asked me to come and jam with the two of them. That's when they asked me to join the band.

We played everywhere that we could. These guys were serious

about working hard and taking the band somewhere. We did parties, clubs, you name it, anywhere that the gas money would take us. After playing all night, I'd basically sleep in my car when I was supposed to be in class. I was getting ready to make the choice—school or the band—when my dad kicked me out of the house.

The Van Halens were just normal guys. We all partied a lot and, being brothers, they fought a lot. They would always hug and make up later, but they would get into these disagreements and we'd have to pull them apart. Didn't matter where—they'd start pounding on each other, but before long they'd be crying and hugging, saying, "I love you, man." Those two had a connection, not just brotherly but musically, too. Ed wanted to hear Al in his monitor. Al wanted to hear Ed in his monitor. They played off each other.

Before Sammy showed up, we were all pretty devastated. It looked like the band had possibly come to the end after Roth left. When we'd first signed with Warner Bros., friends in the business told us that five years was a good life span for a rock group. We figured we may have had our run. The label wasn't very enthusiastic, either. They didn't even want us to continue to call the band Van Halen. Eddie and Al didn't know what to do. They tossed around names of singers and we did have a couple of unknown guys come in and sing with us because we thought that bringing in somebody already known would change the dynamic of the band. That didn't work, and nobody knew what to do until Claudio Zampoli, Eddie's car mechanic, suggested that he call Sammy.

From the first moment I shook his hand on the day he came to Van Halen's 5150 Studios, I knew this guy had a vibe. We had never met before, but I was a big fan: When we'd worked with producer Ted Templeman recording the first Van Halen album, we told him to make us sound like Montrose—we wanted that big "Rock Candy" sound.

Sammy was a breath of fresh air. We went out to the studio

to play him some music. We played and he started singing along. Whatever Eddie could play, he could sing. We all looked at the engineer, Donn Landee, and just went, "Oh, fuck." It was like the clouds cleared, the skies opened up, the sun came out, the birds were singing, the animals were dancing. It was like, "Amen! We've got ourselves a band." He was the perfect missing piece of the puzzle.

We were blown away. We made cassettes and sat there with the engineer. I just remember saying, "We've got ourselves a band." We'd been all down and out and hadn't known what we were going to do. This was the answer to our prayers. This was the kick in the ass we needed. This was . . . fuck, this is it.

I played the cassette for my wife, Sue. It was just the lyric Sammy came up with off the top of his head, which later became "Summer Nights." She went nuts over what she heard. She could tell. You can get together with guys and jam, and I've done that many times in bands, and some of them are great, some of them you just go through the motions. But magic like that is once in a lifetime, if you're lucky.

All of a sudden we're taking it to a new level. Not only do we have this guy who can really sing, but now we've got another guitar player, too. It was something new, something different, and Eddie was really into it. Sammy was the key. He was the guy who took us to a new, higher plateau with Van Halen.

Sam and I hit it off like a steam locomotive. We became friends in a way I had never been with anyone else in the band. The whole band caught the spirit. Nothing was going to get in our way. It was nothing less than a rebirth of Van Halen. There was a lot of energy flowing through that studio when we were working on the *5150* album, ideas coming left and right, all fresh and exciting.

With Sammy, we had real melody. He was just a great all-around musician. Eddie could say, "Hey, Sam, I got this idea," and Sammy could pick up a guitar and go, "Yes, but what about this?" That was all new with us. We started to become a much more musical band.

With Sammy in the band, Van Halen went through the roof. The band scored a string of number one multiplatinum albums. We ruled the arena rock world and played before capacity crowds night after night for more than ten years. We were the world champion hard-rock band and Sammy took us there.

As everything started to fizzle with Sam, I stood by and watched it all happen. I had that comfort zone in Van Halen and I wasn't about to give it up. They were my band.

When Sammy was forced out of the band, and they were working on whatever the deal was for Sammy to leave, I wasn't really a part of those discussions. I just did what was politically correct. I'm sure there were a lot of things that the Van Halen brothers were keeping from me at that point, things that they just went ahead and did.

Eddie Van Halen wanted to be in total control. Al was going along with everything Ed did. Ed took the reins and was just looking for a pawn or a puppet. After Sammy left the band, we didn't really do anything again for years until Gary Cherone joined in 1998.

When Sammy went out on the road with Roth, they thought it was a carnival. Eddie made it clear that he didn't want me to go out with them. I had jammed on some shows before with Sammy and he wasn't pleased about that. But I was just the bass player. My last name's not Van Halen. I didn't feel I was doing anything wrong. I thought I was flying the Van Halen flag. Ed called me when he heard I might appear as a guest on some of the shows. I wasn't going to do the full tour, but Sammy asked me if I'd come out and play a few dates. I was totally into it. I remember Ed saying, "You're not going to be part of that circus, are you?"

When the idea of a Van Halen reunion tour came up, I wasn't in any of those discussions, either. I do know that Eddie didn't want me to be a part of it. I was the traitor because I went over to

Sam's camp. He couldn't understand why I couldn't sit home and do nothing until he decided that we were going to do something.

He wanted to put me on salary, and eventually Sammy, Alex, and even our manager gave up some of their percentage to get me something like 13 percent. It wasn't like I needed the money. The only reason I did that tour was because Sammy was doing it. To do that tour, I signed away pretty much any future rights I might have had to anything Van Halen. If Sammy wasn't there, I wouldn't have even considered it because, by then, I felt like an outsider to the brothers. Somehow I played the shows.

Some nights Eddie would hug me and go, "Mike's back. You're playing so good, man." Other nights onstage, he'd look at me like he was looking right through me, like I wasn't even there. I spent a lot of time keeping my eyes on Alex during those shows, just trying to keep it together. Eddie would say to me, "Look at me all the time."

During rehearsals, Ed would question me about changes in some of the songs. I'd have to change what I played, knowing that was the way it went. The first few rehearsals, Eddie didn't even show up. Alex and I were jamming to all the keyboard songs because we had that on tape. Sam would come in and sing. Everything was sounding good. Then Ed would come in and change it all around. I saw Eddie doing an interview and he actually showed the guy where he marked the neck of his guitar so that he would know where to come in. How we held that together for as many shows as we did, I'll never know.

We'd meet up for sound check midafternoon on our way downstairs at the hotel. Eddie would get in the elevator, ripped-up jeans, no shirt, bottle of wine in his hand. That's where it would start, and he'd take it right through the evening.

At the last show on that tour, Eddie came walking in backstage and he had drawn all over his face and chest with a Sharpie pen. Alex made him wash his face before he went onstage, but the guy was a total fucking wreck. After what happened during the show, I did not even want to fly on the same plane with him. I didn't say good-bye after the show. I never spoke to Eddie Van Halen again. I found out he replaced me in the band with his son, Wolfie, the same time everybody else did—when he gave his big press conference.

After they fired Sammy, I didn't talk with Sam for a long time. Van Halen was my band and I stayed with them. I might have made a couple of comments about Sammy in the press, about his work, but for the most part, I was going along with the brothers. I never slammed Sammy. We had been friends. We were friends right up until when he was fired. After that, if those guys found out that I even spoke to Sammy, that would have been it for me.

It was my wife who encouraged me to go out and play with Sam again. She saw how miserable I was sitting around the house. It was the best therapy I could have had, because I was paranoid about my playing. I might have quit playing bass altogether. After what I went through with those brothers, hell, I was paranoid about everything.

When Sammy and I started talking again, we reconnected on a human level, not really anything that had to do with music. We became even better friends. He helped me out through a couple of rough times. It wasn't just about the music anymore between me and Sammy. We became good friends just for the sake of being friends.

He is the most upbeat, positive guy in the world. He loves life. He only happens to be a singer and play music, too. And another thing—he's no bullshitter. If Sammy says it happened, it did. There's nobody else anywhere like him. I don't know how to put it. Sam is one crazy motherfucker. But I mean that in only the best possible way.

—MICHAEL ANTHONY
DECEMBER 2010

HARD LUCK SON
OF A BITCH

When I was growing up, Fontana, California, was all orange groves, grape vineyards, and chicken ranches. I could eat oranges, grapefruits, and tangerines all fucking day if I wanted. I had to walk through an orange grove just to go to my next-door neighbor's house. There really was no neighborhood. It was long before tract homes. At the corner of each of the long country blocks, there were these big, ten-foot-tall cement tanks with open roofs on them, called water weirs, which fed water to the houses from clean, clear Lytle Creek in the foothills. The water weirs had a float on them, and when the water got too low, the float would kick on and they'd fill back up like a toilet. Each one had a ladder going up so guys could service them, and kids would drown in them all the time. We always heard a rumor that some kid got polio from one of them. But it was our drinking water and it was ice-cold. In the summer, we used to jump in and swim. Not swim, but dunk down, get cooled off, and climb out. I don't want to say we'd piss in them, but we did.

My dad moved to Fontana because he heard the steel mill was hiring. When I was born in Monterey County Hospital in Salinas, California, Dad and Mom had been picking lettuce in the fields and living in a camp where everyone else was Mexican. The Kaiser Steel Mill—the first steel mill west of the Mississippi— pretty much made Fontana. Growing up, every kid in Fontana was just trying to get through high school to get a job at Kaiser Steel. We thought it paid great. Originally, you didn't even need a high school diploma, but, as Kaiser Steel built up, and other plants opened up, making pipes or big beams, you needed a high school diploma. Unless, of course, they got a big order and needed the people. Then they'd hire anybody, and lay you off when they got the job done. But everybody was happy to go there and make whatever they were paying.

It was a brutal job though, working at a steel mill in a 160-degree heat, pieces of hot metal flying out at you. My dad worked in the open hearth, the hottest, hardest work in the plant, where they pour the ingots into big troughs and make steel. Molten fucking steel. He came home with his clothes drenched from sweat, and he used to take salt tablets every day before he went to work. He had probably the lowest job on the totem pole, and they moved his schedule almost every week. He would go from swing shift to graveyard shift to day shift. Sometimes he'd come home at midnight and go to work again at six in the morning. He got burned real bad one time—the side of his face was completely taken off. Just from the heat. It wasn't steel hitting him. It was that he got too close or there was a flare-up or something and it just fucking ripped the skin off his face.

My dad's parents had been migrant farm workers who came out from Kentucky on a covered wagon. They'd picked cotton all the way through Texas, and my dad was born in Texas. Two kids were—that's how long they were in Texas. They had thirteen kids.

He had a younger sister, but he was the youngest boy. My dad was handsome and athletic, but he was a bad little fucker. He would beat the shit out of his big brothers. My uncle told me my dad chased his big brother, my uncle Charlie, up a tree. My dad sat there, smoking a cigarette, waiting for him to come down to beat his butt. Charlie slept in the tree rather than take an ass-kicking from my dad.

My mom, Gladys, was born in Los Angeles. Her dad came over from Italy when he was eleven years old and never learned to speak, read, or write English. He and my grandma—she was Italian, too—never owned a house. They lived in a trailer and were always on the move. He was a chef and he went where the work was. He cooked in Yosemite and went up to Klamath when the salmon were running. He would hunt and fish and work only when he had to. During the winter season, he would cook in Palm Springs, make these huge buffets at the lodge where President Eisenhower stayed. But when the season was over, he'd pack up, take all his money, steal everything he could out of the restaurant, and take off in the middle of the night. The guy was a complete thief—a real crook, my grandpa. And a prick, too. Once in a while he was nice. I'm named after the fucker, Sam Roy. They raised my mom and her sister that way. She grew up in a tent and didn't finish seventh grade.

Mom and Dad got married when she was fifteen. Mom always said all the girls liked him in high school. My dad had dreams. He wanted to be a big-shot kind of guy. He liked hanging around big shots. Bob Hope used to let him caddy on weekends, when he was growing up in Palm Springs. She was sixteen years old when she had my oldest sister, Bobbi. Practically the day she had the baby, as soon as she came home from the hospital, she got pregnant again with my other sister. My sisters Velma and Bobbi are nine months apart.

My father could beat up anybody. I was so proud of that, growing up. He was such a bad-ass. When he was younger, Bobby Hagar fought bantamweight. He won his first eight fights by knockouts.

He was a little guy, five foot eight, same size as me, but that son of a bitch could hit—he could have been something. But he got drafted during World War II, shortly after he'd gotten my mother pregnant again with my brother, Robert. My father shipped out as a paratrooper. He'd never even been in an airplane and suddenly he's jumping out of them. On his first jump, over a battlefield in France, his parachute went way off course. He tangled in trees and smashed his face into a tree trunk. He had a Tommy gun and, as he was coming down, he was scared, so he sprayed the ground with bullets. He banged into the tree and broke his jaw. He cut himself down. He dug a hole, and stayed in a foxhole for a few days. His jaw was killing him. He was disoriented, obviously all banged up from hitting this tree, but he had his gun. Nearby was a German soldier, also separated from his unit, and they played a fox-and-mouse game until my dad killed him in a shootout. I think it really screwed up his head. Killing someone one-on-one isn't like shooting people you don't know. My dad lived with this guy for a couple of days, sneaking around, not sleeping at night, really not wanting to mess with each other, but every now and then, taking a potshot.

When he returned to his company, he was crazy. He was freaked out that he shot the guy. Plus he was a bad-ass anyway. He emptied his magazine in the ground in front of his commanding officer. Told him to dance. That earned him a dishonorable discharge, to say the least, and by the time he came back to California, he was a complete alcoholic and madman. The war had really fucked him up. My mom said when he got home from the war, he used to jump up from bed in the middle of the night and shout, "Where's my Tommy gun? Where's my Tommy gun?"

I was born a few years later, on October 13, 1947, and by that point we were bone fucking poor. But even as I got older, I never knew just how bad off we were. My mom was a great cook and she could make do with things, so we always ate good. I went around hungry a lot because I never had any money. If I wanted to eat, I had to go home and either wait for Mom or cook something myself. I was cooking for myself when I was eight years old. I saw what my mom did. I could boil spaghetti and take canned tomatoes or fresh tomatoes out of our garden. I could make tomato sauce. It didn't seem poor to me. My mom was clean as a pin. Our house was spotless. Our clothes were always laundered. She ironed them, stayed up until four in the morning doing ironing for other people and then ironed our clothes.

My mom always had a chicken coop and we always kept chickens. Whenever we moved (which was a lot), we took the chickens. We never owned a house, and we were always leaving my dad because he was a terrible alcoholic who beat up my mom. When my dad would come home drunk, we'd sneak out of the house in the middle of the night and go sleep in the orange groves. Mom hid blankets wrapped in plastic bags, a flashlight, and little stashes of water and food out in back, ready for the times we had to jump out of the window in the middle of the night.

It was always on payday. He got paid on Thursdays, and when he wouldn't come home directly after work, Mom would begin making plans. He'd come home drunk, start yelling and screaming. He never beat us kids, but he'd thump my mom around. Everybody in the family hated my dad, but they were all scared of him. My sister Bobbi hit him over the head with a baseball bat one time, because he had my mom on the ground. She came up behind him—she was about twelve years old—and bashed him in the head and bloodied up the place bad. My mom got up and we ran. We got out of there.

So we'd leave my dad, and once he was left alone, he would lose the house. He'd stay there, wouldn't go to work, and wouldn't pay the rent, until he'd get kicked out of the house by the cops. He usually got thrown in jail. That was the standard end result of his binge. Sometimes he'd get in the car and get thrown in jail for drunk driving. We'd have to go find a new house every time and move, or my mom would borrow a trailer that her father owned. But somehow we'd always end up with my dad again.

Right before I was born, my mom had a miscarriage. She didn't want to get pregnant. She hated my dad by then. She knew he was crazy and didn't want another kid. She just wanted to raise the kids she already had and get the hell out of the marriage. She had known that for a long time. She had a miscarriage and immediately got pregnant again. She was bumming. She didn't like to tell me that, but later on in life she pretty much told me. "You're lucky to be alive, boy," she said. "If I'd have had that other baby, if it wouldn't have been a miscarriage, I never would have had you." I loved my dad, but he was crazy.

For some reason, my dad was tough on my older brother, Robert. Dad would call him "flea-brain" and my brother would start crying, which only caused Dad to make more fun of him.

"Wahhhhh," he would say. "You sound like a damn siren, you little shit."

He hated my sisters, too. When they turned into teenagers and started seeing boys, that was when the whole thing blew out. He got so drunk he beat up one of my sister's boyfriends. That was the end of the deal for him and my mom.

I was his favorite. I was the king. I was "muscle-brain." He called me Champ, like I was the next champion of the world. He would introduce me to his buddies. "Hey, here's Champ," he would say. "He's got a left hand on him." He was going to make a boxer out of me. Every day, I'd come home from school, and if my dad was there, he'd make me train. He'd make himself a BLT—he was a big BLT man—and sit there in his work clothes, ready to go to work.

"Put on the gloves," he'd say. It wouldn't matter if I'd brought a friend home; my dad would say, "Put on the boxing gloves with your buddy here." He'd make my brother get on his knees to box me. He made me box every day. He'd put on the gloves with me, and teach me. He would take me to gyms and make me hit the heavy bags. "Step into it and twist your body," he would tell me.

My dad was left-handed, so he could pop you totally unexpected, like southpaws can. Even if you know how to fight a little bit, lefties come backward at you. Plus, he was a hard puncher. Some boxers have that gift. There are just guys who can punch. There is something to the magic of timing, how you put your weight, and all these things. Being a southpaw and knowing how to punch, he just knocked people out. He was a one-punch wonder.

Because my dad hit so hard, I learned how not to get hit. By the time I was eight years old, I was getting really fast. I would stand on the outside and move in. He'd try to hit me once in a while and I'd weave. He loved that. He used to really brag it up about me. My brother was bigger than me. He could hit harder. I didn't want to get hit by him, either, so I just kept becoming faster. In and out, in and out. I had a great left jab for a little kid. I used to beat up my neighbors, my buddies. I'd give them bloody noses and my dad would give me a quarter.

Starting when I was four years old, my dad drilled into me that I was going to be the champion. But my mom was practical. "We're going to go pick fruit this week," she would say, "and you're going to go pick raspberries for thirty-five cents a crate"—which meant, at my age, maybe two crates a day—"and you're going to work all summer so that you can buy some shoes for school. Otherwise I can't afford new clothes for you."

I WAS ALWAYS into cars. When I was, like, three years old, I used to stand up on the backseat and lean on the front seat where my mom and dad sat. Going down the road, my dad would point out cars and say, "What's that?" I could name them all—that's a Ford, that's a Chevy, a Studebaker. It was a game we used to play when we would go to San Bernardino to see my grandmother. When I was seven years old, I got a bike, and every October, when the new cars would come out, my pals and I would ride our bikes over to the car dealers a couple miles away to look at all the new cars. There was a Ford dealer, a Chevy dealer, and a Plymouth/Dodge dealer. We'd go around in the parking lots and lift up the hoods and look at the engines. I did that all the way until the Cobras came out at Don Mouff Ford in Rialto. I went over to see the first Mustang when it came out. I was always into it. I'd buy models of all these cars and work on them all day.

I was a straight-A student, the smartest little guy. When I was in fifth grade, before there was PBS, they took a busload of kids to the Los Angeles educational TV station. They only took three kids from my school and filled up the bus with kids from other schools in the district. I was a math genius. You could lay numbers on me and I could do all the math—fractions, decimals, divide, multiply—in my head, just like that. I could go to a world map with no names on it, name the country, name the capital, the river, and could spell everything. Another kid could type eighty-five words a minute or something. Somebody else could do something else. It was a big deal at the time, a reward for only the best students.

I was a hustler, not a thief like my grandpa. I would take our lawnmower and walk around the neighborhood and knock on doors. I had a paper route. I'd ride my bike ten miles to my aunt Maxine's house to wash her car. We didn't have a telephone, so I couldn't call her. I'd just show up.

"Aunt Maxine," I would say, "I need some money. What do you got?"

She'd put me to work. She was real nasty, but she loved me. She was my dad's little sister, so she and my dad grew up as the two babies of the thirteen kids in the same house with one outhouse. She never had kids, and she really took to me. She was a real discipline type. She would work me from ten in the morning until six at night. She would feed me lunch, and then feed me dinner. She would make me take a shower and clean up. She'd wash my clothes, and I would go back home on the bike. She'd give me a dollar. A dollar was a lot of money for me. I would work for that. My brother never did that. My sisters never did that, but I did it. That's why Aunt Maxine loved me. She said on her deathbed, "You were never afraid to make an honest living."

My grandpa might have been a thief, but he was a great chef. He'd take us fishing. He could shoot deer, skin them, and then cut them into steaks in the backyard. He'd make his own wine and can his own food. He'd kill game, catch fish, and always kept a garden next to Grandma's and his trailer. He was always canning, making soups and stocks. Sometimes it seemed to me like my grandpa could do anything. He and Grandma were renaissance people, and they always went back to the same trailer parks. Whenever they were around, I'd go see them. I knew if I stopped by, he would feed me. He could really cook. You could smell his trailer a mile from the trailer park, and he was constantly in the kitchen. That was the Italian way.

But my grandfather could argue, and my grandmother didn't back down. They were always misplacing shit and he'd blame her for it. It was all they'd argue about. There was nothing else to argue about.

"Where'd you put it?"

"I told you, Sam, where I put it. If it's not there, it's not my fault."

"Son of a bitch, then where the hell is it?"

Then they'd get into it. But my grandpa, at the end of every argument—I can hear it in my head today—he'd say, "Hard luck son of a bitch," and that would end the argument. That's all he would say. "Hard luck son of a bitch." I guess my grandma thought, "You know what? You are," and she'd lay off him. I guess they had some hard luck or something, because she bought into it every time.

My grandfather was scared to death of my dad. He had no idea how to stand up to the man. Compared to us, Grandpa had money, but he wouldn't help my mom out. She would have to be broken down on the side of the road, out of gas with four hungry kids in the back, before he'd help, because he was afraid my dad would beat his ass. He had good reason to be afraid. My dad didn't want anybody helping my mom leave him. He was a drunk, and he needed somebody to take care of him. And he would beat anybody's ass, including Grandpa's, if they helped my mom leave him.

One time they were all camping and my dad got drunk and started chasing my mom around the campground. My dad could be something of a sex maniac. When he was drunk, he would come home and want to fuck. My mom would not be into it, because he was violent, and that would just make him pissed off. He would thump her, and probably rape her. My mom wouldn't talk about those kinds of things, but I'd walk in on them all the time in the daytime. I'd open the door and my father would be throwing it down.

So here were my parents and grandparents camping together, when my dad started chasing my mom. But my grandpa thought he was chasing him, which made him start running around the car. My grandma had a few belts in her and she was a feisty Italian lady. She picked up a big rock and tried to hit my dad, but missed and smacked my grandpa right in the face. "Daddy screamed like a woman," Mom always said. Hard luck son of a bitch.

My mom used to really try. She'd go back to him and try, try, try. We'd go find another house. We'd go on welfare half the time. My dad would clean up, go back to work, and pull it together. He would kiss her butt. "I'm so sorry, I'll never do it again." I heard that shit so many times, it was fucking ridiculous. Growing up, that's all I heard. "One more time, I'm leaving you for good."—"I promise you, I'll never do it again." They'd go to church on Sunday for two or three weeks, but soon enough, he'd pull the same shit.

The longest my dad was ever sober was nine months. It was the happiest time of my childhood. We lived in the same house for nine months—an eternity—a nice, big house we rented. My dad made $80 a week. We thought we were living. We bought a brand-new 1956 Mercury station wagon with wood on the side. We actually had Thanksgiving, Christmas, and Easter in the same house with the whole family. Just like every other time though, it all fell apart.

Work only kept his drinking going. He would get in trouble there, but he never lost his job, which helped keep the routine in place. His boss at Fontana Kaiser Steel and another guy at the plant were alcoholics like my dad. They all went to AA meetings together, so he always had his job. These guys would let him come back. He'd go on a drunk for a month and he could come right back to work. That's the way my dad always sobered up; finally one of his buddies would come get him and they'd put him in an AA meeting. My mom would take him back, we'd find a new house, and the cycle would start all over again. We lived in every damn house in town.

As I grew older, it got kind of thin. People knew. You would go over to somebody's house and—small town—my dad would have already been in a fight in a bar with the kid's father, which basically meant that I wasn't welcome.

The same went for girlfriends. I had this one girlfriend, Pat, when I was in eighth grade. She was my first love—eighth grade, slow dancing. I didn't have a car, so I'd walk her home from school and then I'd walk home, about three miles each direction. Her parents wouldn't let me come in the house because her father had gotten in a fight with my dad in some bar. I thought they were rich. He was a contractor or something and had a little bread. They lived in a nice, big tract home. One day her mom took pity on me and invited me in the house. She knew I was a nice kid. I stayed for dinner, and I was so fucking uncomfortable it was ridiculous. We went in the den and slow-danced and made out. It was getting late when her dad came in to tell me I had to leave. I started to walk out and Pat started in on her father, "Oh, Daddy, give him a ride home." They had a brand-new 1960 Thunderbird. What a car. I got in that car with that guy, and he didn't say one word. I didn't want him to go to the house, so I had him drop me off on the corner—"This is good, let me out here." He burned rubber getting out of there, not to show off. He just wanted to get rid of me. It made me feel really bad. And I wasn't a bad kid at all. Not yet.

FONTANA WAS COMPLETELY segregated. If you looked at the geography of the town, Sierra Avenue ran from one end of the town to the other, right down the middle. Route 66 ran through Fontana—that was Foothill. The next street up was Baseline, and black people had to stay on the other side of that. Down at the southern tip of Fontana was Valley. Mexicans had to live down there. If you came down into the little shopping area in town, the cops would harass you. White people would beat up black people. I saw it with my own eyes. If a black guy was walking down the street, a carload of white guys would pull over and beat his ass.

My dad never had a black friend, I'll tell you that. No black guy ever came back to our house. If my dad ever worked next to a black guy, if any black guy ever worked alongside whites at Kaiser, he never told anybody. The "n" word was prevalent in my house.

My dad never laid a hand on one of his kids, ever, except one time with my brother. My mom was ironing, and my dad came home from work at four in the afternoon. He was covered in black soot. We were usually in school and didn't see him come home from work. My brother shot his mouth off. "Look at Daddy," he said. "He looks like a nigger." Dad ripped the cord out of the iron and beat my poor fucking eight-year-old brother with an ironing cord. Mom had to pull him off.

My dad could be uncontrollable. Back when he was boxing, he'd been suspended from fighting because he attacked the referee who wanted to stop the fight. After he wasn't a fighter anymore and no longer had his license, he would go to the fights with my uncle Cleo, who was married to my mom's sister. And—this is how screwed up my family was—Uncle Cleo was also my dad's nephew, even though he was the same age as my dad. It was uncle-sister-brother-aunt-and-nephew all in one. Uncle Cleo loved my dad. They were asshole buddies from day one and they would go to the fights on the border in Calexico and Mexicali.

"Your dad, hell, man, he wasn't afraid of nobody," Uncle Cleo told me.

They'd be watching the fights, passing the whiskey back and forth, and there would be a couple of quick knockouts. Dad knew the fights were going to end early unless they could make a couple more matches, so he would go backstage, drunk as hell, and do some shadow-boxing. The promoter would say okay and he'd go in there—just take off his shirt and street shoes, put on some boxing gloves, and fight. My dad weighed, like, 135 pounds, a few pounds heavier than his fighting weight, and he'd fight guys around 175 pounds. He would go in there and get his head beat in. He would keep getting up. He wasn't even in shape. He'd just go in there swinging. He'd get five bucks.

My father didn't have much of a fight career, but it turned out he did write himself into the boxing record books. Years later, I was watching Tommy Hearns fight for the light heavyweight championship and he kept knocking the other guy down. I was living in Mill Valley, watching TV. One of the commentators said, "That's got to be the record for the most knockdowns." Somebody obviously went and looked it up, and a couple of minutes later, the announcer said, "No, the record is held by Manuel Ortiz, who knocked down Bobby Hagar twenty times." How fucked up is that?

He would beat up the neighbors. He would beat up his brothers. Any time we had a family reunion on Thanksgiving, everything was fine until afterward. He and his seven brothers, their nieces and nephews, they'd all start playing pinochle. You'd hear, "Oh, you son of a bitch! Boom." Shit would start flying. The women would run out of the room. "Get the kids out of here, Robert just jumped Leroy" or "Robert's fighting Carl." It would take all his brothers to hold him down and chill him out. That's the way he was. Fucking crazy. I dug it. I watched through the window. He was such a bad-ass.

It may sound like my dad was a bastard, but it's not that simple. He was the weirdest guy, and when he wasn't picking fights, he also had this big, soft heart—at least when he wasn't drinking. He would sit there and look at the mountains or the ocean and say, "Isn't that beautiful?" When he was sober, this tough, crazy bastard would say those kinds of things. I used to always think that was strange. I thought, "Wow, my dad's soft." It would rub me wrong when I would hear him say something was pretty or he would act real loving with me. I thought he was like a big, tough guy, and now he's acting like a sissy. I didn't know how to relate to that. It turned out he was real sensitive, too, but he never had a chance to show it. Back where he came from, the sensitive guy got his ass kicked. He had to be tough. But I think he wasn't that way at all. I think he was like an artist. I got it from somewhere. He was a good singer. He used to sing to the radio. He'd yodel his ass off. He was a trip.

Mom didn't leave him for good until I was about ten years old. It didn't happen all at once, and it started with a car. Somehow, between picking fruit and berries with us kids, ironing and who knows what else, my mom bought this old '36 Ford for thirty bucks, and she hid it on the other side of town. My dad didn't know. Then, when we left him, we could sleep in the car, instead of the orange groves. My mom would drive to orange groves and park in the middle somewhere where cops couldn't see you. Once in a while the police would come, but they weren't pissed off. They would just say, "Lady, this is dangerous, you sleeping here with these kids. Please go park in a neighborhood or something." That's how personal the whole thing was in Fontana in those days.

The final straw came when my dad set the house on fire. It was following a gray year when Mom had taken a real stand and rented a house on her own. It was this fucked-up old house, basically a chicken coop somebody converted, and it was only a block and a half away from where Dad lived. But that was the first time she said, "I'm staying here—we're done with this guy." A lot really happened in that gray year. She took a real job—went to night school to learn how to type and landed a job as a shipping clerk at a hosiery factory. Instead of ironing or picking fruit, she would work all day, pick us up from school, and take us to the field, where we would all pick fruit or whatever until sundown.

For a while there, it was back and forth, back and forth between the two houses, but then Dad fell off the wagon after a long time sober. It was payday. My mother came back from shopping with the money and he was gone. Uh-oh. He came home in the middle of the night. We scrambled out the window. Sometimes she would let him fall asleep and she would climb back in the window once he conked out. She'd spy on him and, when she was sure he'd passed out, she'd bring us back inside so we could sleep in our beds instead of sleeping in the cold outside. We were living in this two-story house, and my dad fell asleep in the bedroom on the second floor while he had a lit cigarette. By the time we came home to see if he'd passed out, we saw the smoke and went back in the orange groves. The fire department showed up and ran upstairs carrying hoses. My father woke up and started kicking everybody's asses. They had to put the fire hoses on him before they could put out the fire. The cops came and took him to jail.

He stayed in the burned-out house. That was the last of his having a house. Bit by bit, my mom would come in during the day, or when he passed out, and take our stuff out of there. This time, she was gone for good. She was taking furniture. She took everything that we could use, everything that was ours. We were living in this chicken coop and my brother would walk to school past the old burned house. It only lasted about a month before my dad finally hit the streets. They threw him out of the house. But the last time my brother saw him there, he snuck up to the window and peeked in the living room without my dad knowing. He said Dad was parked on a wooden crate, sitting in that empty room in the burned-out house with a bottle of whiskey in one hand and a cigarette in the other.

✢2✢

MOBILE HOME BLUES

Ed Mattson taught me how to play guitar and drive a car.

Ed was three years older and had gone to school with my brother. In high school, Ed had been this fat kid with a big nose. Everybody made fun of him and beat him up. His mother was a slumlord. She had about ten houses in bad neighborhoods. She had three on the street where they lived in Fontana. I never met his father, who was a steelworker in Gary, Indiana. They didn't live together, and, after meeting his mother, I could understand why. His mom was a Russian Jew and completely crazy.

Then one day, Ed got a nose job in Hollywood, went on a diet, and lost tons of weight. His mother bought him a brand-new 1962 Chevy Super Sport Impala. Suddenly he was the coolest guy in town and nobody knew who he was. My brother didn't recognize him. He had this fabulous pompadour. He was going to Hollywood and having his hair razor-cut by Jay Sebring, one of the guys that Charles Manson would eventually kill. His mom gave him money and he always wore cool clothes. I used to see him driving around, usually by himself. He picked me up hitchhiking one time, and we became fast friends. He taught me how to drive. I had a learner's permit and he let me drive everywhere. He took me to Sebring's salon and got my hair cut. He thought I was a young guy with some potential and he wanted to help bring it out.

He played guitar, and, early on, he was listening to Bob Dylan. Ed turned me on to the Beatles and the Stones. Before that, Elvis had been my first real hero, because my big sisters loved him so much. When I was a little kid, my sisters would have parties. They would dress me up, ducktail me out, and every one of those girls would dance with me. I dug the hell out of it. I was a little hard-on, a nine-year-old kid with these girls five or six years older, starting to get sexy, starting to get little titties. All because of Elvis.

Even though I dug Elvis, I didn't pursue music. But the Beatles got to me. I was in high school, already had a girlfriend, but I still couldn't help liking "I Want to Hold Your Hand." When I heard the Stones, I was gone. Local Top 40 deejay George Babcock brought the Rolling Stones to San Bernardino in June 1964 for the band's first concert in America, and I fucking went. Ed Mattson and I drove over in his car to the Swing Auditorium with the plan to sneak in. We stood at the back door when along comes this clunky, fucked-up orange school bus, and George Babcock gets out of the bus with the Stones. They walked in the back door, and we walked right in behind them.

That was when it started for me. That night, I knew I wanted to be a musician. Ed Mattson could already play guitar. I started singing with him, and we would play these Beatles and Stones songs. We knew three or four songs. Ed said we needed two guitars, so he could play lead. By that time, my mom had a job and was doing okay, so I tried to con her into buying me a guitar on her Sears account. She told me that if I learned how to play "Never on Sunday," the popular foreign film title track played on a bouzouki, on Ed's guitar, she would buy me a guitar. I learned it in, like, a day or two. She bought it on time. It was $39.95—a Silvertone guitar and an amplifier in a case.

My first band never played a gig, but we had capes. A couple of my pals and I helped paint a house for one of their fathers. We used the money to buy the fabric and talked one of the guys' mothers into making them—black velvet, red lining, Dracula collars. We wore them around town one day and people driving down the street would honk their horns and flip us off. We thought we were really cool. We decided we would wear them to the Dick Dale dance in Riverside, but we were too young to drive. We hitch-hiked. Nobody was going to pick us up wearing those capes, so we rolled them up and put them on again when we got to the dance. Only they wouldn't let us wear the capes into the dance, so we rolled them up again and stashed them outside in the bushes. When we came out, they were gone. Somebody stole them. My mother thought the whole thing was hilarious, and she never let me forget it. Years later, when she was interviewed for the local paper about me, she told them, "Oh, his first band—they had capes."

Dick Dale was the thing. The King of the Surf Guitar moved his weekend dances from the Rendezvous and Aragon Ballrooms on the coast to the Inland Empire, and his weekly shows at the River-side National Guard Armory were big events. When Ed Mattson wanted to start a band, we called it the Fabulous Castilles and played surf music. I learned Dick Dale's "Miserlou" on guitar. We never had a drummer in that band, and we all played through one amp with the microphones. We practiced every day.

We met Jerry Martin, a fifties oldies-but-goodies, doo-wop dude, but a good bass player. He was way above the level that Ed and I were playing at, but he thought we knew what was going on and would let us come over. I had started having long hair. I started wearing rock-and-roll-looking clothing. To us, he had it made. He had a fifties band that would play nightclubs all over the Inland Empire and L.A. He had his own house, drove a new T-Bird, and even had a single out that they had played on the radio one time. He would play with us in his garage because he thought we were hip and onto the latest things. Soon he started growing his hair and getting hip himself, but I was still astonished when he pulled out a joint and said, "Hey, want to get high?" I smoked a little pin joint with him, but didn't feel anything. I was too nervous. It took a couple times.

Once I discovered rock-and-roll and pussy, I barely made it through high school. Especially after I got my first guitar, I was done. I was flunking everything. My teachers would say, "Why don't you apply yourself?" That was high school. All I cared about was music and girls.

And once I finally got high, it was really all over. One of my pals and I took a joint and drove way down the dirt road in the Jurupa Hills, almost to Riverside. I couldn't stop laughing. I couldn't drive. I couldn't do shit. I dug it. After I'd decided I was going to smoke dope, in order to get that dope, you had to find the lowlifes of all lowlifes. These were the dope dealers in Fontana—not hippies but bad-guy heroin dealers—those were the people you went to see if you wanted to buy some weed. It wasn't even good weed. So there I was, becoming a stoner, smoking dope every day, taking acid, even shooting speed. Shot heroin—didn't dig it.

After high school, I wanted to get out of Fontana as quickly as I could. Fontana was a blue-collar town. Nobody would have ambitions like being a doctor, a lawyer, or anything like that. You were either a bad-ass, low-rider Hell's Angel–type tough guy, or you got a regular job and got married immediately after high school. There were no plans on leaving that town. My father was the town drunk. He was living on the streets. I'd be driving my car around and there was my dad, stumbling down the sidewalk, drunk. Every so often, we'd bring him over to my sister Bobbi's house and she would clean him up, but it would never last.

Mom was doing much better. Not only did she have her own real job, she met a guy, Mike Majerovwych. He was a big Russian bear who escaped the Ukraine after the Soviet army slaughtered his family while Mike hid in the basement with his baby. He ran away on foot and gave the baby to a couple in a village on his way. He made it to Canada, where he worked in a coal mine, and eventually moved to California. He was living in Cucamonga and working as a chef at Chafee College when he met my mom at a polka dance. He was a good guy, although he didn't like my long hair and didn't mind letting me know. Still, he bought me my first car.

I was getting ready to make my jump. I was aching to be a musician, but I worked at ABC Stores in the automotive department. I wore my hair in a Bob Dylan Afro and had this girlfriend, Christie Carson, who looked like Twiggy. I started living with her at her grandmother's house in San Bernardino. I bought a brand-new black 1967 Volkswagen—$1,900 out the door minus $300 trade-in—that my mom cosigned for and I was making payments. I talked the manager of the ABC Store into opening a music department at the Riverside store and I ran it. I ordered all the records. I started getting record players. I started stocking guitars. I built the whole music department. But I was doing drugs. I was smoking dope constantly. I was stoned all the time. I'd wake up in the morning and smoke roaches before I went to work.

My friend Bucky was really into music, and he liked to play. He was a year older than me, but during the summer, we spent time together. I was always hanging around with tough guys and Bucky was kind of a tough guy. I'd go over to Bucky's house. He had a swimming pool. I thought they were wealthy. Sometimes I would see his younger sister, Betsy, sitting around in a bikini. I knew her from high school, although she would never say hello to me in the halls. I was always tripping on her, but Bucky would say, "Don't mess with my sister—I'll kick your ass."

Bucky would come over to ABC Store and buy these cardboard record-filing boxes that sold for $1.39. Little, cheap folders with A-B-C-D-etc. on them and you put your records in there. Only I would fill them up with new albums by Cream, Jimi Hendrix, all the stuff we wanted to hear, charge him for the filing box, put everything in a paper bag, and staple the receipt for $1.39 to the bag for the security guard to check on the way out of the store. Sometimes I would slip in a new album by Donovan or Joan Baez for his sister. So this one time, Bucky's got the bag with about twenty pounds of records in it, and the security guy goes to grab it. The bag ripped to pieces, the box hit the ground, and all these records went flying all over the place.

They busted Buck. I got fired. They wanted to fire me anyway. They were watching me and knew I was smoking dope during the lunch hour. But that wasn't all bad. I went on unemployment. I was getting high all the time and playing guitar.

I knew this small-time dope dealer in Fontana named Jim, and one day he asked me to drive him to San Francisco to score some LSD and go to this big rock festival in June 1967. I had the car. We drove up to the Haight-Ashbury, and arrived about four in the morning. We crashed in the car on the street. When we got up, Jim knocked on the door and we went into this apartment full of speed freaks. I watched as some guy shot speed in his fucking neck. I was not digging the scene, but Jim scored a bag of windowpane acid. We dropped the acid, bright and early in the morning, and drove down to the Monterey Pop Festival stoned out of our minds.

The concert took place inside a fenced-in arena, surrounded by a park. Thousands of people were hanging around outside. I spent most of the time out in the park outside the arena. Everyone was sitting under the trees, taking drugs, smoking pot, burning incense, having sex in the grass. We saw Brian Jones wandering around, as stoned as we were. We saw some of the bands. We caught Otis Redding. I saw Eric Burdon and the New Animals, who blew my mind. Hugh Masekela was cool. I saw weird things like the Association, who had great harmonies, but I missed Hendrix. I didn't see Janis Joplin, either. I was high on acid for three days.

All that left a big impression on me, but it was when Bucky introduced me to *Fresh Cream* that I decided that's what I was going to do. I hadn't had a job in a bit, and I'd been spending my time getting high every day. I'd been listening to the Paul Butterfield Blues Band's *East-West,* and of course everybody dug *Sergeant Pepper's,* but not really trying to play anything, just tripping. But then Cream came out with their first album, and all of a sudden, I've got to get back into music, way back. I wasn't just tripping anymore. I went and stole a guitar—actually a friend of mine stole it. He knew about the guitar and where it was. We watched and waited for the people to leave their house and we climbed through their back window. My friend did it because he believed I was going to be a rock star. I already looked like one. Another one of my doper friends, Tim Tameko, who had a real job at Kaiser Steel, signed for me to buy an amplifier, a Fender Bassman, because that's what Eric Clapton used to get a big, fat tone. I set my sights on becoming Eric Clapton. I went to see Cream play that September at the Whisky a Go-Go on the Sunset Strip and I thought I made eye contact with Eric Clapton. I thought he looked right at me. I looked just like him. I wore my hair in a big natural Afro like his. It was before he had his long, wavy hair. I had the exact same look.

THE NEXT MONTH, I went over to see this band I had heard about. They already had a lead singer who played guitar and he was pretty good, and I only went to check them out. I walked in wearing a pinstriped double-breasted jacket with a roach clip on my lapel, white T-shirt, jeans, boots, John Lennon shades, and my frizzy Afro hairdo. The band's guitarist, who was named Jesse Llamas, took one look at me and said, "Can you sing?" Just like that, I was the new singer for the Mobile Home Blues Band.

Jesse and I were pretty much opposites. He was this fat, greasy-looking Mexican slob who didn't even tie his shoes but could really play, while I looked the part but had little else. Jesse knew what was happening and turned me on to Jeff Beck. He worked at this cool little record store between Colton and San Bernardino, where he bought an import single of "Rock My Plimsoul" and, on the other side, "Hi-Ho Silver Lining." Jesse played me "Red House" and "Manic Depression" off the English version of the first Jimi Hendrix album.

The Mobile Home Blues Band was a weird band. We only did ten gigs or something, but we rehearsed for what seemed like forever. The highlight, if you could call it that, was when we opened for American Breed (remember "Bend Me Shape Me"?) at this crazy, nonalcoholic club called Purple Haze in Riverside. Our bass player, Benny Mosteller, was a little genius in a wheelchair. He was born with this rare disease, all crippled up. His joints were fused together. When he was about fourteen years old, he had an operation and they just cut out the joints, so at least he could straighten out. He played a Hofner bass, like Paul McCartney. He was right-handed, but he had to play left-handed because of the operation, and he was fucking great. He was a brain, too. He would tell everybody what key and what notes to play. He had long hair and he was really fucked-up-looking from spending his life in a wheelchair. But he was a math genius in school. And he played perfect—perfect time, perfect note choices.

His mother was the greatest woman. She was a single mom,

and Benny had a totally normal brother, too. His mom would feed us and let us rehearse in their garage. She would pick Benny up, take him out of the wheelchair when he had to go to the bathroom. She would wipe his ass for him and everything. She was the sweetest woman and she loved us because we didn't care that he was all fucked up. She put up with us bringing girls over all the time. The drummer was a fourteen-year-old kid who didn't even have his first drum set yet. His name was David Lauser, but Jesse called him Bro. Forty years later, he is still my drummer. We had a rhythm guitar player and, the first time we played a gig, he freaked out. He was shaking so badly he couldn't play. He had to play behind his amp. He couldn't even come out onstage. And we kept him in the band. I was just the singer in the band. I was doing too many drugs. I'd gotten in the habit of going over to Tim Tameko's house to smoke pot. Every now and then I'd buy a lid and deal joints. I'd go over to Tim's house and act like I was cool, because I was trying to be a rock star now. It was a scene. There was always music going. I'd play guitar and he'd let me leave my amp there. He would always have joints or acid or speed, and people were constantly coming and going. Guys would be shooting up speed. I came in one night late. Everybody was smoking dope. I had a couple of joints. I lit one and passed it around. This one guy said, "Hey, man, do you mind if I have this to go?" A roach? I didn't think anything of it. "No, it's cool."

This random guy turned out to be a narc. They were trying to get the big dealer, whoever that was. There was no big dealer in our world. We were all dealing joints, not kilos. If I had $85, I would have bought a kilo in a second, but I never had eighty-five bucks. Still I didn't realize what I'd done until about three months later, when a cop pulled me over and gave me a ticket for an illegal lane change. It was such bullshit that I went to court to fight it. When the judge asked me how I pleaded, I said, "Not guilty."

"And how do you plead to the charges of possession . . . no, distribution of marijuana?" he said.

"I don't know what you're talking about," I said. I had no idea what he was talking about, and then this narc walks in holding a plastic bag with my joint. It turned out that our pal Jim had been busted and rolled over on everybody else. They threw me in jail.

In the middle of the night, some guy in my cell pulls a baggie with two joints out of his asshole and fires up right there in jail. It freaked me out. "I'm in for this, you motherfucker," I said.

"They don't care," he said, and I knew he was right. We toked up. Nothing made sense.

It was not a good trip. I stopped smoking dope after that, because getting high in jail bummed me out so bad. There was some enormous bastard in the next cell for writing bad checks. "You got a girlfriend?" he says. "She's fucking Jody right now, man. He's fucking her in the ass. She's blowing him." Some other old jail guys chimed in. They were dogging me heavy and they were getting to me.

It was the worst four days of my life. My mom was bumming hard. My brother and my sister visited me. Everybody's going, "Oh, Sammy, how could you do this? When you get out, you're going to cut your hair and go straight," and all this shit. When I finally went to court, I didn't have an attorney, so they gave me a public defender. When we eventually did appear in court, it turned out that I'd worked with the bailiff at the ABC Store where he had been working as a security guard. He recognized me, took the judge aside, and told him I was a nice kid, a hard worker, there must be some mistake. The judge let me off. It was a lucky break. I could have done time.

So I got out of jail and I decided no more dope—I'm playing music. I'm getting rid of all those friends, quitting all that shit around Tim Tameko's house. I stopped making payments on my amp because I didn't have any money, and there was still $80 left to go. Tim paid it off. I went in, like, a year later and they told me it had been taken care of. I never saw Tim again. He got busted, too, only he went to prison. Me? I never worked another day in my life.

❊ 3 ❊

GOING TO SAN FRANCISCO

Johnny Fortune saw me with the Mobile Home Blues Band at the Purple Haze. He was famous in the area for his hit single, "Soul Surfer," that made him something of a grade-B Dick Dale for a minute, but his shelf date had actually expired years before. Still, he was a local name, and in 1968, he approached me about being in his band, agreeing to pay me a hefty $150 a week to let me sing whatever I wanted. I wore a mustache, sideburns, and grew my hair long. It was a funny act—he'd play some twangy, old-fashioned rock-and-roll guitar instrumental and I'd sing something from Cream or Hendrix—but we were working steady at the Club Tyro in downtown San Bernardino.

One night, Bucky's girlfriend took a couple of fake IDs and came in with his sister, Betsy. I brought them sloe gin fizzes on the break because it seemed like the sort of drink for a girl. She might have had two. Betsy got dizzy and I took her outside and held her in my arms. She'd never had a cocktail before in her life. She'd never used drugs. Standing there, holding her, I kind of fell in love with her. I was still living with Christie at her grandmother's house, but I started hanging out with Betsy and ended it with Christie soon after. It wasn't long before Betsy and I took a camping trip to Big Sur and decided to get married. I was twenty-one years old when we were married on November 3, 1968.

I got sick of the Top 40 scene with Johnny Fortune pretty quickly. I just sang the same stuff, and it got old fast. I started hanging out at this dump of a nightclub in nearby Riverside called the Gasser, where the few original rock bands there were in the area played, and I ended up meeting this guy named Dave Arney, who was a bit older. He was a bass player, and he wanted to start a real band—an original band. He already had a drummer, Larry Taylor, who also wanted to work on original material.

This fit with where I was. I didn't want to play covers and neither did they. We started a band called Cotton. Betsy sometimes sang background vocals and played flute. Arney had a van and he had connections. He knew an agent who got us a gig in San Francisco, backing up oldie-but-goodie acts like the Coasters, the Drifters, the Shirelles, even Bo Diddley, who was very cool.

When Betsy and I got married, this was our honeymoon. We found hotel rooms above the Basin Street West on Broadway. Betsy and I had our own little room and the other guys shared a room. The bathroom was down the hall. I wasn't a junkie, but I decided I was done with my drug days and tried to clean up. I didn't even want to smoke dope anymore because Betsy didn't dig it. I'd get stoned and get bummed out, so I did my best to cut it out altogether.

This wasn't easy, considering there was always a lot of weed around. Pounds of it. Larry Taylor, the drummer, had never taken acid before we got to San Francisco, but sure enough not long after we got there he took acid and went wandering out in the street. He finds two cops walking the beat on Broadway and says, "You guys want to get high?" They say, "Sure, we'd like to get stoned." He brings the two cops, four o'clock in the morning, back to our house—thank God Betsy and I had our own room. They fucking arrested the fuck out of Larry and Dave.

So that band, obviously, was over. I had nowhere to go. I had no money. I was married. I went back and stayed with Mom for about five days. My stepdad, Mike, crucified me. Every morning he dragged me out of bed and told me to find a job. The last straw was when Mike butchered a hog in the front yard and brought the head in to show everybody. Betsy freaked out. She ran off and wouldn't come home for hours. I couldn't take it anymore. Betsy's dad had a trucking company with three dump trucks in Rochester, New York. He didn't collect garbage, but he hired his trucks out to the city. Betsy's brother, my pal and new brother-in-law, Bucky, was driving one.

"Betsy, call your dad," I said. "We've got to get out of here—I'll do anything."

Bucky's truck-driving was a pretty new job for him. A while back, Betsy and Bucky's dad had moved back east, leaving her and Bucky to take care of their house in Fontana. Unfortunately, Bucky became a heroin addict, and after Betsy and I got married and split for San Francisco, Bucky sold everything. They had a '59 Corvette. He sold it. He stayed in that house with no water or electricity. He started breaking into houses, robbing people. He was really desperate. His dad got wind of everything, came out from New York, beat the shit out of Bucky, threw him in the back of a car, and drove back to Rochester. Bucky kicked in the backseat of the car on the way to New York but that didn't stop his father, and by the time Betsy and I were ready to leave California, Bucky was already back there driving one of his old man's trucks.

I didn't have any money, and around then, my Volkswagen blew up. Betsy's dad really didn't like me, but he loved his daughter, and he gave us a pair of tickets for a Greyhound bus that stopped in every fucking town between California and New York. It took us three and a half days to get to Rochester. I had an amplifier and a guitar, and we had a trunk with all of our stuff. We'd change buses a lot of places. You'd get to Chicago and have to get on a different bus and I'd lug everything. Betsy couldn't carry anything. She was pregnant and wasn't feeling well the whole time. We didn't have any money. We didn't eat on the bus. Betsy's dad picked us up at the station. He didn't even look at me.

I started driving trucks for him. Every day of the week, four o'clock in the morning, we picked up junk and hauled it to the dump. Each truck needed three guys. We'd pick up casual laborers, black guys waiting on a street corner, to hang on the truck, jump off, and pick up the stuff. I was one of those, only I came automatically with Bucky's truck. I made $10 a day.

I was looking to make money any way I could. I was in a band called Salt and Pepper with this guy named Herb Gross. When Herb originally asked me to be in his band, I told him I needed money, so he got me another job, at H. H. Sullivan Printing Company. I got off work with the dump truck at eight o'clock in the morning, ate breakfast, and then went to work at H. H. Sullivan until five o'clock in the afternoon. If there was any overtime to work, I'd do that, too. At the same time, I rehearsed with Salt and Pepper every night. We never played a gig, but we rehearsed the whole time I was there. In the middle of all this, I got drafted, and because Betsy was pregnant, I sent in her medical report. They didn't draft guys who had children. That was the summer of Woodstock. It was happening just across the state and I wanted to go, but Betsy was too pregnant.

I worked all that summer. I saved every penny and didn't spend one nickel. We ate for free at the house. I didn't buy a record. I didn't buy a guitar string. I didn't buy a candy bar. I always told Betsy's family I wanted to go back to California and I wasn't messing around. Betsy's dad built this van for us to drive back to

California in. He put a brand-new engine and transmission in this old, beat-up van, and some nice, new tires. We worked on it to-gether—him, Bucky, and me. I had two jobs, was in a band, and I was working on the car on weekends. The first time I started that car up, I was ready to go. Bucky threw our shit in the back and, at the last minute, jumped in with Betsy and me. He didn't even say good-bye to his parents. His dad wanted to kill him. It was four o'clock in the October afternoon, the first day of snow that year in Rochester. I had my money. The second that engine turned over, we fucking split.

Because the van only had two seats in the front, we put a little cot in the back for Betsy, so she could lie down, nothing more than a patio lounge chair that we wired to the side so it wouldn't slide around. Bucky and I took turns driving, and we went straight through in fifty-six hours. Ran out of gas once and broke down once, but Bucky was a good mechanic and he fixed some wire on the engine. We got it going again. We ran out of gas somewhere in the middle of nowhere. He hitchhiked, got some gas, came back, and we kept going. We did not stop.

When I came back to Fontana, I had saved about sixteen hun-dred bucks. My sister Bobbi's husband, James, had rebuilt my Volkswagen for me, and Betsy and I rented a nice little house on Anastasia Street. I had enough money to last a few months, but I still didn't have a job, so I went on welfare. I couldn't afford to have the baby; food stamps and welfare saved my ass. Now I didn't need to have a second job. I could live and play music. That was when I started a new band called Big Bang. It was with Bro, the drummer from Mobile Home Blues Band, David Lauser. The year before, I'd seen the Jeff Beck Group with Rod Stewart at the Fillmore West in San Francisco and more than ever I knew what I wanted to do: play guitar like Beck and sing like Rod. There was an old garage in the back of the house on Anastasia Street where we rehearsed. We ran through a bunch of bass players and other musicians and went through a number of names—Manhole, Skinny, even, for a brief minute, Chickenfoot.

Aaron was born February 24, 1970. When she finally had the baby, Betsy went crazy almost immediately. She had a nervous breakdown and experienced panic attacks that made her unable to breathe. She was really in trouble, and she ended up in Ward B in San Bernardino County Hospital. She had to see a psychiatrist.

While Betsy was struggling, my sister Bobbi took care of little Aaron, and somehow I barely noticed what was going on around me. My wife's having a nervous breakdown. She's twenty-one years old, just had a baby, and somehow I still had a one-track mind. Nothing else—not my wife and not my new son—mattered to me as much as playing music. I was playing music with Big Bang every day and trying to gig at night. We had a gig here, a gig there. We played an all-black club in Riverside called the Hat Factory, six sets a night starting a ten o'clock, ending around six in the morning. We won a Battle of the Bands playing "Manic Depression." Got a trophy and a teardrop Vox electric guitar that we traded in for some public-address gear. We finally settled on a bass player, a boyish-looking college student named Jeff Nicholson.

Finally, Betsy's psychiatrist wanted to see me or we faced getting thrown off welfare. He needed to find out what was going on. I told the guy I was going to be a big rock star. He called me Peter Pan. "You need to wake up, son, and go out and get a job," he told me. "You've got a wife and a child." He really laid into me. That just pissed me off. I was ready to take him outside. I was a firm believer in myself. I was going to make it. But I might have looked like a total asshole, with my long hair, bad attitude, no job, on welfare, a baby, my wife in a hospital. When Betsy came out, my sister Bobbi nursed her back to health and helped her out a lot. I didn't change one bit.

Luckily, though, a steady job was not that far off. I was driving in my van with my equipment in the back, going to rehearsal in San Bernardino in somebody's garage, when I saw a guy boarding up the windows on a club out by Highland Park called the King of Hearts. It used to be a real nightclub, for the martini crowd. I said, "Hey, are you open?"

"Nah, we're shutting down," he said.

"I've got a band," I said. He was kind of listening to me. "Can we play here? We'll play for free."

The couple who owned the place had no money. They were only barely keeping it open. We started playing and the first night two people came in. Next weekend, we had twenty people. Pretty soon, we were drawing a crowd, a hundred people in there, jam-packed. They changed the name of the place to The Nightclub. We started charging twenty-five cents at the door. We kept the money and then he started paying us. We added an incredible gui-tarist named Bob Anglin and a keyboard player, Al Shane, and we changed our name, one last time, to the Justice Brothers. We worked four or five nights a week, pulling down a mighty $25 per man every night. That was all cash, so I was still getting welfare. I was living on Easy Street.

I WAS LYING in bed one night at the Anastasia Street place in Fontana, asleep, dreaming. I saw a ship and two creatures inside of this ship. I couldn't see their faces. I just knew that there were two intelligent creatures, sitting up in a craft in the Lytle Creek forest area about twelve miles away in the foothills above Fontana. And they were connected to me, tapped into my mind through some kind of mysterious wireless connection. I was kind of waking up. They said, in their communication to each other, no words spoken, "Oh. He's waking up. We've got to go." They fired off a numerical code, but it was not of our numerical system. There was a split second where I was still seeing everything, and then it was over, like someone pulled the cord or whatever.

I opened my eyes real quick. My whole room was white. I couldn't see anything. No fixtures, no nothing. It was a timeless white. Infinity. I couldn't move. My eyes were open, but I was paralyzed in my bed. Betsy was lying next to me. All of a sudden, *pow,* the connection instantly broke.

I jolted. The room went back to black. Everything returned to normal. It was four o'clock in the morning. I was shaking. My heart was pounding. I was scared out of my brain, beyond anything I ever experienced before. What was that? I didn't even tell Betsy. That set me off on the weirdest quest. I didn't even know the word "UFO." I didn't know my astrological sign. I didn't know anything about astronomy or numerology or anything. But I dug into it.

I started looking up in the night sky, sifting through my dreams more often, looking for patterns, breaking things down, and reading books. In the back of the yard was an old, abandoned chicken coop. It wasn't even on the property, but next to the driveway, a dilapidated shed with a roof ready to collapse. One day I decided to check it out and see what was inside. The door came off in my hands. Inside there was nothing, except for a dirty, fucked-up trunk. I opened it and the only thing in the trunk was a book on numerology. I've always been a bit of a mathematician. I started reading and it tripped me out that if you add numbers up, you always come down to one number. You can take, say, 137. 7 + 3 is 10, plus 1, that's 11 and that's 2. Or you can stop on the master numbers: 11, 22, 33, 44. Numerology is like astrology. It's just mathematical equations. I got that by thumbing through the book.

Then I started getting deep into it. I'd add the numbers in my home address to see what it was as a one-digit number. I discovered that if you add 9 to anything, it disappears. 9 + 1 is 10. Back to 1. 9 + 7 is 16 or 7 again. Any numbers divisible by 9 always comes back to 9. Three 9s are 27 and 7 + 2 is 9. It will always come back to nine. Four times 9? 36. 3 + 6 = 9. Whenever you add a nine to anything else, it disappears. That intrigued the hell out of me. It drove me crazy. I went, okay, if you added 999 to 9,999, it's 9 again. You can go on around the block with nines. It always adds up to 9. But if you add 9 to anything else, it disappears. I not only read the entire book, but many others on the subject. I became a numerological nut.

In my family, you didn't talk about psychics or astrology or stuff like that, but there was a lady named Miss Kellerman, who lived in a tract home in Yucaipa. She didn't take reservations. You just pulled up in front of her house, from eight o'clock in the morning to whenever she felt like shutting the door. She had a screen door in front of the regular door, and if the regular door was shut, it meant she wasn't seeing people. If that door was open, it meant she was doing business. You'd knock on the door and give her fifty cents. She'd take a look at you, and if she didn't want to see you, she'd just shut the door.

I tried to go see her probably ten times in my life, and there was always a line of cars down the block. But one afternoon after I'd discovered the numerology book, I pulled up there with my fifty cents, which was about all I had total. The door was shut. There were no cars. I drove all the way out there—I was going to go knock on the door. I knocked. I had a little bit of a beard, really long hair, stone hippie to the bone. She opened up. She barely cracked the door, put her head out. She was an Italian lady who spoke in broken English. "I'm not seeing anybody," she said.

"I drove all the way out here," I told her.

She took a real hard look and stared right through me. Then she opened the door. I started following her back to this little room, she started talking. "You need to shave your beard, but don't cut your hair," she said.

For an old-time Italian woman like her to say don't cut your hair in those days was amazing. All my mother wanted me to do was get my hair cut. "Don't cut your hair," Miss Kellerman said. "It looks very good under lights."

We went into her back room. She sat down in a rocking chair and put her hand under a velvet cloth that was on the table. She began fidgeting back and forth in the rocking chair with her eyes closed and just started laying it on me.

"Don't take drugs," she said. "Who's Bill?" she asked. "Who's Bill?" I didn't know.

"Don't be mean to your stepfather," she said. "He loves you very much and he's going to die soon." She had it right, only backward—it was my father who died soon.

"You have a brand-new little baby girl," she said. "No, it's a boy but everyone thinks he's a girl."

That was Aaron. He was born with long, thick, and curly black hair. People always thought he was a girl.

"He's a beautiful little boy," she continued. "And your wife, she has a real problem with her breathing. She needs to drink raw egg whites with honey and lemon. That will help her breathing."

Even though Betsy had left the psych ward, she continued to have panic attacks where she couldn't breathe. We had to rush her to the hospital in the middle of the night, like an asthma kind of thing.

"You used to smoke cigarettes," she said. "Don't start smoking cigarettes again. It's bad for your voice."

Don't do this, don't do that. Don't take drugs. Don't cut your hair. That was the one that got me, since my hair kept me from getting jobs, kept me from getting in places.

"You're going to move to Northern California," she said. I'm what? I'm playing with the Justice Brothers in The Nightclub. I've got no money. I'm on welfare. I just got married and have a brand-new kid.

"You're going to move to San Francisco," she said, "but you need to go to Santa Barbara first. There's something there for you. But then you're going to move to San Francisco and you're going to make it. I see your name in lights all over the world. And you're going to go to seven countries. You're going to go to Italy, France, Germany . . ." She named off all these countries and I'd barely been outside Fontana.

"Go to San Francisco," she said. "That's for you."

And I believed her.

I had wanted to go back to San Francisco ever since Cotton broke up. I had seen the scene and it blew my mind. I went back to Fontana to put together a new band so I could move to San Francisco. That's what I was trying to do in New York. I went all the way to New York, where I knew I had a job, just so I could make some money to get back to San Francisco. That was my goal the whole time. "Start a band," I thought, "and if we can make it in San Francisco, we can make it anywhere." Because it's a hard-core community, I felt the pressure. It was all about music and wasn't bullshit. It wasn't L.A. I hated the L.A. scene. I wasn't about to jump through that hoop. I was going to San Francisco, where people could be themselves. I wanted to have my own band that would be exactly who we were—like Moby Grape and all those bands we loved that came out of San Francisco.

I went back to Fontana and told my band to pack. On the way, we drove through Santa Barbara. We stopped at a little college place outside town. They had food, but it was a hippie kind of place. "My band and I are on our way to San Francisco," I told the guy running the place. "Can we play?"

"I can't pay you," he said. We told him we would just pass the hat and he said okay. We set up and played around five o'clock in the afternoon, right when dinner started. A few people came in. They didn't give us much.

After that stopover in Santa Barbara, we continued on our way. We didn't move to San Francisco full-time on that trip, but it was the first of many. We would drive up on our nights off and audition at clubs. We landed a job playing Monday and Tuesday at the Peppermint Tree on Broadway in San Francisco. We won the gig by playing "Won't Get Fooled Again" and "Behind Blue Eyes" by the Who. The manager had never heard a club band do that before. We would load up our gear on Sunday nights after playing all weekend in San Bernardino and drive up. The eight-hour drive was never fun. The only way all of us could fit in the van with the equipment was if somebody lay across the top of the Hammond B-3 organ in the back. The Peppermint Tree was a slick place. The bands that played there had to be good. They wore uniforms. They were show bands. But Monday and Tuesday were the off nights. We would play for nobody. We played covers, but the Who and the Stones, not the commercial rock hits. But we also played our originals.

Eventually, the San Bernardino scene at The Nightclub got out of hand. The Hell's Angels started coming around and they tore the place apart. They would start kicking ass and the fucking National Guard would have to be called in. We'd go out and sit in the back. Finally the city just shut it all down. We were devastated. But when The Nightclub got knocked out, I said fuck it. We're going to live on those two nights in San Francisco.

When we got to San Francisco in fall 1970, at first, we all slept on the floor at the apartment of Don Pruitt, who I knew from my previous time in San Francisco when he was the manager of Basin Street West. We finally landed a job at a club called the Wharf

Rat, not a hip joint, but the Justice Brothers was a pretty good band and it was a steady gig. That gig, along with welfare and food stamps, helped me make enough money to get Betsy and Aaron off Pruitt's floor. The band continued to live with Pruitt at 1565 Oak Street, and Betsy and I rented our own little, one-room flat in the same building. Aaron slept on a mattress under the kitchen table, and Betsy and I stayed in the living room with a mattress on the floor.

One night, I dreamed somebody was knocking at the door. I got up to answer it, wondering who could be knocking on my door in the middle of the night. I open the door and it's my dad. Only he's, like, twenty-two years old, young and vibrant. "Hey, son, great day for the Irish!" he says. He's acting crazed, really happy, but drunk on his ass.

"What the fuck are you doing here?" I tell him. "My kid's sleeping right here on the floor, right in this room. Don't ever come here drunk in front of my family. You're going to scare this guy. Now get the fuck out of here." I slammed the door shut, went back to bed, and that was how the dream ended.

Two minutes later, somebody's knocking on my door again. I head to the door, thinking, "God damn it—I'm really going to chew his ass out this time," and I opened the door. It's Don Pruitt from next door. "Your sister is on the phone," he said. I didn't have a phone of my own. I went over and picked up the receiver. "Dad died," my sister said.

Dad died in the backseat of a police car. They picked him up in a park in San Bernardino across the street from The Nightclub. He had been living on the streets since we left him in that burned-out house. After the cops had picked him up, they'd arrested three more drunks on the way downtown and all the drunks started fighting in the backseat. One of the cops sprayed Mace over the idiots and when they arrived at the police station and got out of the back of the car, Dad just fell over and, for once, didn't get back up. I often wonder what would have happened if, in my dream, I'd have invited him in instead of yelling at him.

Once we all had enough money to afford apartments, we left Oak Street and moved into two flats in the same building in Noe Valley. The band had the upstairs and Betsy and I were downstairs. Those guys would be upstairs with Nicholson's stereo set, staying up all night, while I was downstairs with Betsy and Aaron. Betsy had a broomstick she would bang on the ceiling. Bob Anglin, the genius guitarist, would fall asleep smoking cigarettes and leave his record player on repeat.

Even before I ever heard his music, I saw a picture of David Bowie and I immediately knew. This was one cool-looking dude. This cat's got it going on. When I heard the music, I also instantly became a big Mick Ronson fan. He had that killer guitar sound. I wanted to be Bowie and Ronson just like I wanted to be both Jeff Beck and Rod Stewart or Mick and Keith. I was up there singing James Brown songs dressed like Bowie. I used to put glitter on my chest, wear makeup, lipstick, eyeliner—the works. Betsy started making my clothes, and I wore satin pants with these big gay-ass boots. She was really good. I would spend all my extra money on material. We'd go to the fabric store and find the most far-out satins and silks. She made me velvet pants. I was getting really out there, and it all looked pretty gay. I'm the last guy on the planet who'd seem gay, but for some reason I really dug that. We played a couple of gay parties, but I was wondering what was up with that. I'm dressed like a woman. I'm wearing high heels and makeup and I'm wondering why gay guys are digging it?

We were packing the Wharf Rat, every night, and making good money, $175 a week, real money, especially for a rock band doing cover tunes. I was writing originals. I started getting more outgoing onstage, becoming the Bowie. Jeff Nicholson hated me being the leader. They all wanted to be funk musicians, but I wanted to be more like the Stones. I was always into soul and blues, that's how I came up. The only middle ground for the Justice Brothers was doing Tower of Power. But I got tired of learning other people's songs. The last song we learned was "Can't You Hear Me Knocking" by the Stones. Nicholson sang it because I wouldn't do it. I walked out of rehearsal and they started rehearsing without me.

✤ 4 ✤

MONTROSE

Twelve o'clock in the afternoon, in a full-blown David Bowie outfit—high heels, makeup, glitter, nine yards—I drove over to Sausalito in my beat-up Chevy van with my Les Paul to meet Ronnie Montrose.

I'd first seen the flashy guitarist only a week or so before, when I went to catch the Edgar Winter Band at Winterland in San Francisco in spring 1973 with the rest of the guys from the Justice Brothers, who were there to see Tower of Power, also on the bill. It was a rare night off from the Wharf Rat, and I was all glittered up. I wanted to see Edgar Winter, because he was glitter rock. That was all I needed to know. I didn't know who Ronnie Montrose was, even though he'd already been recording with Van Morrison, but I'd seen him on TV and I dug his moves. He had this little thing where he crouched down with his Les Paul and he spun in a circle, leaning on one foot. He went around and around. He didn't get tangled up in his cord. It was a pretty good move. I was impressed.

The next day I'd started hammering the guys in the band, telling them that's the kind of guitar player we should have. We already weren't getting along that good, but everybody wanted to keep the steady job. I'd talked about it with John Blakeley, one of the few guys around town I knew from Riverside, who was in a band called Stoneground, but I didn't even know Winter's guitarist's name.

"That's Ronnie Montrose," Blakeley said. "He lives in Sausalito. That was his last show with Edgar Winter. He's looking for a singer."

And so that was how I ended up knocking on Montrose's door wearing a silver suit and boots. I brought four songs I had written that the Justice Brothers wouldn't do. He plugged me into this little amp he had and I played him "Bad Motor Scooter," "Make It Last," "I Don't Want It," and "One Thing on My Mind." He played me the riff to "Rock the Nation"—that's all he really had. I showed him the lyrics to "Space Station Number 5" and he started playing that riff. We wrote the song together that day.

I thought he was rich. From where I stood, he seemed like he was in the biggest band in the world. They'd sold out arenas and had got a number-one album. What did I know? I didn't know the house was a rental. I saw his car outside, a '63 Ford four-door. I thought he was driving a pretty beat-up car, but that didn't faze me. All I knew was that he had just banked an $8,000 royalty check—a fortune, in my eyes.

When Ronnie came to see me at the Wharf Rat, the Justice Brothers were pissed off, but they kissed his ass. He walked in all rock-starred-out in a crushed-velvet jacket, big rings on his fingers. After the set, we went outside and he said, "Let's start a band." I quit the Justice Brothers that night. One week later, Nicholson was wearing my exact outfit. He got Betsy to make him the clothes. He's wearing the glitter, the makeup, my whole getup. He's doing my act.

Ronnie asked if I knew any drummers. A while back, I'd sung on a demo tape by this band, Thunderstick. They were very much like Free, totally Paul Rodgers, not very Northern California–sounding at all. They were very English glitter-rock, but they didn't look it. They were all in jeans and T-shirts. I had the look. They wanted me bad. They didn't have a record deal but they had a record-company guy who was interested but didn't like their singer. I did some demos with them and I was thinking I might go with them, but nobody in the Justice Brothers wanted to lose that Wharf Rat gig, not even me. Then I saw Edgar Winter at Winterland. Denny Carmassi was the Thunderstick drummer. Denny got the gig, even though we tried out the great British rock drummer Aynsley Dunbar. Ronnie didn't want any competition. Ronnie wanted control, so he wanted guys like me, who didn't know anything. We sat down at Studio Instrument Rentals and auditioned a number of bassists, including Andy Fraser of Free, who turned out to be a complete junkie and never even showed up. We tried Ross Valory, before he joined Journey, and Pete Sears of Jefferson Starship. Ronnie knew Bill Church from the Van Morrison band and, before that, a little blip of a band, called Sawbuck, that played on some Fillmore bills. Edgar Winter's drummer, Chuck Ruff, also belonged to Sawbuck. Church sat there the whole time, watching us try out all these guys. I kept telling Ronnie I liked Church.

"Yeah, but the guy's kind of an asshole," he said.

I thought he was great. Every time we'd take a break, Church and I would go outside. "Ronnie's such an asshole," he said. "He knows I can play. He knows I should be in this band." Ronnie didn't want to use Church, because they had fucked around on each other's old ladies or something like that. Of course, we ended up with him. But Ronnie tortured him for about a week, auditioning other guys.

Ronnie knew Ted Templeman of Warner Bros. Records through Van Morrison. Ted came up and saw us. We rehearsed at Studio Instrument Rental maybe three or four times and we had the whole first album written. We wanted to be Deep Purple or Led Zeppelin.

"Rock Candy" was the last song we wrote. We had nine songs, one called "Drugs" and another one called "We're Flying," which weren't very good, and we threw those out. Templeman told us we needed one more good song. Denny just started playing that drum beat when Ted was in the room. Ronnie came up with the riff and I just started singing, "You're rock candy, baby." The song just came together. That was the only song that was a band song. All the rest of the songs, either Ronnie or I wrote.

Ted signed us. We went straight into the studio with him and engineer Donn Landee. Everything happened so fast. We moved into the Sheraton Burbank, near the Warner Bros. lot, where all the acts stayed. We had no money. They gave us a $50,000 advance, but we spent $25,000 on equipment. We each took five grand. We kept five grand in the bank. We didn't have a manager. Ronnie was totally in charge.

I got my $5,000. I rented a house for $80 a month in Mill Valley, 37 Montford Street, and I bought a car. Not just any car, of course, but a Citroën Deux Chaveux, the most uncool car on the planet—a French car that looks like a sardine can. I thought it had class. I sold my VW to a guy for fifty bucks. The van was so bad that, when I sold it to somebody, it couldn't make it out of the driveway. It was too steep. The guy had to back up in the dirt and get a running start, because it couldn't make it. It was that powerless. The Citroën cost almost three grand. I rented the house and had, like, $1,200 in the bank. I was rich. We made a little session pay making the album. I now had a phone. I got my first credit card. I knew the album was going to come out and I had written those songs. One cool thing Ronnie did for me was that he had our lawyer set up my own publishing company, Big Bang Music, so that I could control my publishing rights.

We were going on tour, so we started interviewing managers. We met with the guys who handled Loggins & Messina. When we met Shep Gordon, who managed Alice Cooper, he was wearing a sarong and sandals and a ratty old T-shirt. At the time, Alice Cooper was a really big star, but, as it turned out, hadn't made a lot of money. Shep explained all this to us. They spent their money making Alice a big star. All the promotion, all the hype, all the marketing.

"Do you guys want to be rich or famous?" he said. "You've kind of got to sacrifice. If you want to be famous like Alice Cooper, you've got to spend a lot of your money on things—big production, big publicists, the big image thing. On the other hand, take a big band like the Doobie Brothers—they're so much richer than Alice Cooper, it's ridiculous."

I'd been in Ted's office and seen an $800,000 royalty check for the Doobie Brothers, who had "Listen to the Music" at that time, which put me on the fucking moon. Ronnie didn't like Shep. He was too loose and weird for Ronnie, plus he had too many of his own ideas.

Ronnie wanted Dee Anthony, because he handled the J. Geils Band, and Ronnie had toured with them back in the Edgar Winter days. So Dee Anthony became our manager. That was big-time, because he had J. Geils; Humble Pie; Spooky Tooth; Emerson, Lake and Palmer; Joe Cocker. He had the roster. So we just went out to tour and basically never came home.

Every night we were opening for either Humble Pie, J. Geils, or Peter Frampton—all Dee Anthony acts—and then Black Oak Arkansas and Foghat. We'd get in a station wagon and drive ourselves. We didn't have a tour manager, so Ronnie had put us on $150 a week salary when we started to tour, $10 per diem on the road. He was the one to check us into hotels, and he was a great band leader, at the start.

A lot of the time, we opened for Humble Pie. We didn't go over that great, but we were working or traveling seven nights a week. Sometimes we'd even do a club the same night. We'd open for Humble Pie at the arena, then run over and do an eleven o'clock show at the club to make a little extra money. We had two roadies. They drove the truck with the equipment, and when they got to the place, they'd set it up. Then, during the show, one guy was our lighting guy and the other guy was our stage manager.

We played Detroit, like, every month. We opened for everybody and their dog. About the twelfth time we were playing Detroit, we were opening for Aerosmith at Cobo Hall. They were big there, but nowhere else. It was right after their first album. We got a huge encore. Montrose was really starting to get big in Detroit, and we came back to do "Helter Skelter," the Beatles song. It was one of two or three encores we used on our first tour. We would go out and play the first Montrose album, come back and play a cover. "Helter Skelter" was one of them. We came offstage, go back to the dressing room, can't wait for these guys we've never heard before go on. Denny and I ran out to the side of the stage. They went out and opened with "Helter Skelter." They didn't know we'd done it.

Opening for all these different bands was a crash course in touring. Humble Pie spent every penny they made on tour. They'd fly Lear jets. Steve Marriott, their lead singer, would come off tour with nothing. They were selling out arenas everywhere. But they lived so high, they were always broke. Steve was such a fuckup. I loved the guy. He and Peter Wolf from J. Geils were guys that I watched every night when I opened for them. They taught me how to be a front man. I remember one time in Chattanooga, Tennessee. We were sitting in the hotel room of tour accountant Jerry Berg, picking up our $10 per diems on Monday morning, first in line for the weekly payout. Jerry was filling out the paper, sign here, when Steve came busting into his room, fucked up in the middle of the day. He'd been up all night doing blow and drinking.

"How much fucking money we got, mate?" he said.

Jerry started to close his briefcase and Steve punched him in the mouth, grabbed the briefcase, and dashed out the door. Carmassi and I just sat there, stunned that we didn't get our per diems. Jerry was bleeding. Steve was gone. He had a limo parked out front.

Berg got on the phone with Dee Anthony, who wanted to know how much money was taken. When Berg told him forty thousand dollars and change, Anthony went nuts, chewing him out. As far as he was concerned, it was all Berg's fault. Marriott didn't show up for the concert that night. We had to cancel and now we were stuck in Tennessee. The next day, pulling into the Holiday Inn parking lot, here comes fucking Dee Anthony, rolling in like the president, practically with little flags on the car. He gets out, grabs Jerry by the neck, throws him against the wall, and gives him another pounding.

They found Steve in jail. He was arrested with a bunch of drugs, hanging out with some black dudes. He was a soulful little white British boy who wanted to be black and sang like it. They got him out of jail and the tour continued.

About a month later, Steve came up to me in a hotel lobby and said, "Hey, mate, let me borrow your cassette player." I had a little bullshit cassette player and headphones and used to walk around with it. Everybody did.

"Uh, no, not really, man," I said. He invited me up to his room anyway.

I went to his room. He's got a suite. He's got a framed picture sitting on his table, covered with a road map of cocaine. At least an ounce. Half of it gone, big chunks missing. He's got a couple of chicks and a couple of other guys hanging out.

"Put your tape recorder down," he said. "Let's fucking play

some music." He pulled out a blues cassette, John Lee Hooker, something like that. I did a couple of bumps. I'd done coke twice in my life. The first time, I didn't even realize I'd done it. I didn't feel anything. I didn't realize I was fucking numb, that my whole face was numb. The shit was that good.

"Hey, man, you got like a hundred-dollar bill or something?" he said.

I pulled a five-dollar bill out of my pocket. He rolled it up, did his blow. I did mine. I see my five bucks sitting there and I'm thinking, "Before I leave, I got to get that." That was the first time I'd ever hung with Steve. He was cool. He didn't give a fuck. He was completely on the moon all the time. He'd stay up until he passed out. He'd do blow until he ran out. He'd spend all of his money, and when he was broke, he'd go earn some more. He'd just live like that, really a reckless guy, but cool. Before long, I had to go. I couldn't take the crazy coke thing. I had to get my tape recorder and my five bucks and get the fuck out.

I took the five bucks, unrolled it, and stuck it in my pocket. I went to grab the tape recorder and Steve piped up, "Oh, no, man, you've got to leave the tape recorder," he said. I knew if I left my tape recorder, I'll never see the thing again. But sure enough, I left it. I actually did get it back, like, a week or two later, from some equipment guy who cleaned out his room, but that was the last time I partied with Steve. It just didn't feel good in there.

One night we were off in Atlanta. Montrose used to play this club, Poor Richard's, but this time they had a blues band with Willie Dixon, the guy that wrote all those Muddy Waters and Howlin' Wolf songs. Steve had a limo. I waited for him to get offstage. We jumped in his limo and he poured me a Courvoisier and handed me a quaalude. I didn't drink and I didn't do drugs, but off we go to the club. Steve wanted to jam. We walked in, sat down at a table. I started getting dizzy. I was fucked up. Steve went up and talked to the band. They said okay and he looked at me. "Come on, man," he said.

He jumped up onstage. Richard's had a stage that was about two feet high, right about knee level. I went walking up. I walked right into the fucking stage and fell flat on my face and passed out. Next thing I know, I woke up in my hotel room, going, "What the fuck happened?"

THESE WERE DAYS of growth and learning for me. I was searching in life. I was reading the Alan Watts book *This Is It*. I was reading books by the mystic mathematician Ouspensky—*A New Model of the Universe, The Fourth Way,* and *Tertium Organum*. I was reading Einstein's Theory of Relativity. I was totally uneducated. I was in a rock-and-roll band, one of the lowest things on the planet, but I had these great big ideas. I wanted something and I couldn't help thinking about it.

One night I freaked out Ronnie. Very seldom did Ronnie and I share a room. This night we found ourselves in the same hotel room in Atlanta, Georgia, and Ronnie was anxious about the band. "This isn't really working for us," he said. "We're not making it. We're in the hole. We're losing money. I don't know what we're going to do. What's your idea? What do you want to do?"

I laid the whole thing on him that I laid on Nicholson and the boys. I was glittered up in Montrose. I was shiny. I was satin, velvet, rhinestones, and platform boots. Betsy was still tightening me up with my clothes, making really nice things.

"I think we need to really step it up," I said. I told him we should get really shiny, put on a big show. "Let's get some backing from Dee Anthony or Warner Bros., and throw this big production together. Like Alice Cooper—let's put on a show. Then when we go out and open for these bands, we'll blow them off the stage and we'll make it."

I went on for probably an hour, and then I heard a *click,* the light went out. He didn't even respond. I'd laid the whole thing on him, stuff that I'd been thinking about for months. And he turned out the light. The next day, everything had changed. It was over. He wouldn't even include me in the conversations with the other band members.

Ronnie wanted to be a rough-and-tough jeans and T-shirt guy, don't talk to the audience, never smile. He had a lot of anger inside of him. He could have been in Metallica or something like that today. He shut me out. I started feeling real insecure. At sound check, he would play with Denny and not be concerned about what I wanted to do. He was hurting my feelings.

The first Montrose record never made the charts. It made Bubbling Under one week, but it never graduated to the actual *Billboard* album charts. By the time the tour was over, we had sold eighty thousand records. I went into Dee Anthony's office in New York and said, "Hey, Dee, Ted Templeman told me we've sold eighty thousand records." He picked up the phone and calls up the booking agency—I'm sitting there, watching him do this—and he said, "Frank, these guys sold eighty thousand records. Let's start asking for seven-fifty, maybe a thousand dollars."

It just sold slowly, but it sold. We didn't have a big Top 40 hit when we came out, but FM was picking up "Space Station" and "Rock Candy." It's still never been on the charts, but the first Montrose album has sold more than 4 million records over the years. "Rock Candy" is like a standard for bands like Def Leppard or the Cult. Over the years, anybody who wants to jam with me wants to jam "Rock Candy"—Chad Smith, Joe Satriani, Matt Sorum, Slash. Lemmy from Motörhead came up to me at a show in England, and what did Lemmy say? "Fucking 'Rock Candy,' mate."

It is in the bible—it's scripture.

By the time we came off that tour, we had all sorts of money issues. About a month into the tour, all that per diem stuff became sporadic, and when we came out of the first tour, we were owed ten weeks' back pay. Dee Anthony had helped us out a little bit, but not much. We ran out of money on the road. We got stranded in a Holiday Inn in Little Rock, Arkansas, and we couldn't get out because Ronnie's card was maxed. They called the cops and made us sit in the fucking hotel. We called Dee Anthony and he finally gave them a credit card over the phone.

We were making $500 a night and it was costing about $600 a night to tour. We were dying. Without getting paid, my home phone was shut off. Betsy was back on Montfort Street in Mill Valley with Aaron, sitting there freaking out, not able to talk to me. It had made the latter part of the tour miserable.

When we came off the road to do a second record, Denny and I went into this little studio I had in the basement and wrote a bunch of songs. I wrote "Call My Name," "Someone Out There," and a handful of songs that ended up on my first solo album, *Nine on a Ten Scale*. I wrote them for Montrose, but Ronnie didn't want to even hear them. He wanted producer Templeman to find outside material. We cowrote three songs. "Paper Money" was some of my best lyrics yet, a little political commentary on consumer society. Even Templeman wondered what Ronnie was doing. He insisted on being named coproducer with Templeman and he was in the booth, saying he wanted things to sound this way and that way.

After that conversation in Atlanta, he totally shut me out, held me back, and pushed me down. He did not want me taking over that band. I didn't want to run his band, but I was looking to make it. I thought I had a great idea for that band. It might not have been the right idea. I can't say Ronnie was totally wrong. But he got so insecure about it that he broke up a great writing relationship. He could have nurtured me. Those were my first songs.

Looking back at my life, I can't call anything a mistake. I've had nothing but great success, and really in a nice chronological pacing that brought me to where I am today. If I'd experienced huge success in Montrose, I wouldn't be here today, doing this. It's all been little steps that have kept opening my mind. I would have stopped growing a long time ago if it hadn't been for Ronnie and guys like him. Knowing that makes it hard to be pissed at him now. But, at the time, I was going, "That motherfucker."

While all this was going on with Montrose, my marriage was really on the rocks. My wife, Betsy, was losing it. In spite of her breakdown after Aaron's birth, I didn't realize just how serious her psychological problems were. She was incredibly needy and I liked the way that made me feel. The first time I'd smoked dope around her, she couldn't handle it, but I'd held her and felt very compassionate, like I could help this person. That made me feel very manly. But then, when I was on the road, Betsy would not let me get off the phone until she fell asleep. I'd have to sit there for hours after a show. She'd say things like "I took Aaron over to Lori's house. I had to take Aaron over there because I'm depressed and I'm freaking out and I'm afraid." I thought she might kill herself.

Lori was Denny Carmassi's wife. They didn't have kids, but Lori was a real solid lady, very strong. I dealt with this every night. I would be on the phone. I would spend every penny I had on hotel phone bills. I cared about her. I was concerned about my kid. But I was focused elsewhere. I was determined that I was going to stay in that band. I was going to make it happen. That was very tough for me. She made life hard for me.

I sent Betsy down to stay with my sister Bobbi, who'd helped her after Aaron was born. My sister practically raised Aaron, because Betsy was always depressed and crying. She would get to the point where she couldn't get out of bed. When I came home, she'd be happy, but soon she would be accusing me of fucking around, and start her "When are you going to get out of this business?" line of questioning. I told her when I have a hundred thousand dollars in the bank, I'll quit, enough money to start another business.

When the band started doing a little better, I told her she could come out on the road with me. Ronnie had a firm rule against wives on tour, but I told him I didn't have a choice. The first time we went to England, she borrowed money from her parents and flew to England on her own. Aaron stayed with my sister Bobbi. Betsy traveled by train and met up with me in hotels and didn't even go to the gigs. She busted her ass to be with me. She was insanely jealous. A girl would look at me and she'd turn beet-red and attack her. For such a humble, meek little mouse, she could be very aggressive. If a good-looking woman walked in front of me with Betsy, I'd have to look down. She'd watch my eyes, everything I did. My childhood sweetheart, my virgin bride, who I never gave reason to worry. Not at first, anyway.

It was rough, because Montrose basically went out for two and a half years and never came back. We would come home for a couple of days or a week, then head down to L.A. into the recording studio, and then back out on the road.

That's when that whole promiscuity thing with me started, and it just got worse and worse through my time with Montrose. I was trying to be good to Betsy. I'd feel guilty if I was with a chick. I wouldn't sleep with them overnight. Mostly I would just let them give me a blow job. I didn't think that was cheating. I thought that as long as I didn't fuck them, that's not having an affair. Then I decided as long as I don't fuck them twice, it's not an affair.

I had all these little rules and I was trying to be a good guy.

I was probably the best guy in the world for about two years. In those days, at every gig, outside of the hotel, hanging around backstage, in the dressing room, on the side of the stage, everywhere you went in our little world, there were groupies. They were dolled up and they were there for the band. Whatever you want. You want to get your dick sucked? You want to see them eat each other's pussy? You want to fuck them? You want to take them on tour? That's what they were there for. Some of them good-looking, some of them not, but always dressed up to the tits with a lot of makeup. They shined themselves up good.

It was so available. I'm a sexual person, so coming off a stage, being up there, playing the rock star for thirty-five minutes, if I saw a woman that really tweaked me, I couldn't resist. It was not an easy thing for me to deal with. I think it destroyed my marriage more than anything. I had to lie so much and I'm not a liar. I'd rather tell you the truth and deal with it. Otherwise you wind up with so many things to deal with that when you see that person coming, your stomach starts churning.

The Warner Bros. Music Tour was the beginning of the end. Montrose, Bonnaroo, Little Feat, Graham Central Station, the Doobie Brothers, and Tower of Power—three bands each night, two nights in every city, all across Europe in February 1975. We went by train and we played everywhere. Denny Carmassi and I had never been out of the country.

One of Don Pruitt's buddies used to tell me, "Motherfuckers don't wear clothes on the beach in Europe," and I would always tell Denny this story. So when we finally arrived, we get off the plane and check into our Holiday Inn in Munich, and we go, "Let's go down to the pool." We wanted to see the naked chicks. The hotel had a spa, with an indoor/outdoor pool and a restaurant above. We went in the men's locker room, dropped our drawers, grabbed our towels, wrapped them around us, and headed out for the pool. We went through the turnstile, unwrapped the towels, and went to jump in. We're naked. Both of us. Some German guy started yelling, *"Nein! Nein! Nein!"* We were walking, buck naked, turned around and looked up. People having lunch in the restaurant were looking down at us. They ran us out of there hard. The guy chewed us out in German. We didn't know what the fuck he was saying, but he was definitely going off on us. I felt like the dumbest asshole that ever lived. I went back to my room and wouldn't leave.

In Europe, Ronnie wouldn't talk to me. Montrose headlined some shows. Other nights, we would open for the other bands, like the Doobie Brothers. We opened for Little Feat in Amsterdam and got booed off the stage by the third song. People started whistling in the middle of our songs. That really destroyed Ronnie. He decided it was over. I could see it in his head. He wanted to break up the band. I knew it. I heard him talking to Denny.

"Heavy metal's done," he said. "We're in the wrong kind of band. We need to get out of this."

Warner Bros. was paying us, like, $250 a week and all expenses. I was making more money than I'd ever made yet in Montrose. I got this great review in Belgium. It had a picture of me in the newspaper the next morning. It was in Dutch, but it was a positive review about me, what a great front man, a new star, and all that bullshit. That was it for Ronnie.

We went to Paris the next day. On the trip, I got food poisoning from some mussels I ate in Belgium and was violently sick, puking and shitting. We had sold out two nights at the Olympia Theater—Montrose was big in France—and it was the last tango in Paris, the final night of the Warner Bros. Music Tour. We pulled up outside the theater. An old pal from the Humble Pie road crew, Mick Brigden, was driving us. Ronnie was in the front seat. I was sitting in back with Denny and Alan Fitzgerald, our second bass player, because Ronnie had fired Church before the second album. Ronnie turned around to talk to me.

"After tonight, I'm quitting the band. What are you going to do?" he said.

I was sick, edgy, and ready to punch the guy in the face. "I'm going to start another fucking band. What the fuck do you think I'm going to do?"

He reached back and shook my hand. "Okay, good luck," he said. He didn't even look at me onstage. I was sick. I couldn't sing. I was weak. It was a horrible ending.

The next day, on the plane on the way home, I talked to Denny about starting another band. A week later, Ronnie called him and the other guys and told them he was going to keep Montrose together and get a new singer. It was totally premeditated. He had it all figured out. Our record deal was up, and he had already told Ted Templeman that he wanted to renew the record contract and that he was getting a new singer. They were going to re-sign for a lot more money.

I got home from that tour with nothing in my pocket, no money in the bank. My wife was freaking out and I had nothing going on, nothing coming up. I didn't think I could make my next month's rent. But almost immediately, a publishing royalty check from the first Montrose album, for $5,100, showed up in the mail. I had no record deal. I had no way to make a living, except for playing music. I knew I was going to be okay, but I didn't have it set in stone. So I went out and I bought a $5,000 Porsche.

5

THE RED ROCKER

As soon as I got home from being fired, I walked into my sister's house and went straight to the phone. Betsy had come over to Europe again and Bobbi had been taking care of Aaron. I went straight from the airport to her house. I picked up the phone and called Dee Anthony and told him Ronnie fired me. "Hold on," he said. "Don't be dropping no bomb on me. I've got to look into this. You guys owe me a lot of money." I never heard from Dee again.

Jerry Berg, who was Dee Anthony's tour manager, always liked me. When I told him I needed a manager, he quit his job to manage me. He was a first-class guy, smart, always well dressed. He handled the money and I thought he had his business down. He seemed like a good business guy, but he really didn't know what he was doing managing a band. Dee Anthony had been doing everything and keeping Jerry in the dark.

I was on my last nickel. I went over and stood in line and started collecting unemployment again. I was considering going back on welfare, too, because I had a baby. My house rent in Mill Valley was more than $200 a month, which, in those days, was a lot of money—way more than I was able to make in a month. I was fucking down and out, brother. People in San Francisco thought I was a big star. They figured these guys headlined Winterland. I would have thought we'd made it, too. But I didn't have any money.

I took an old mattress that I picked up off the side of the road and put it up against the wall in my basement for soundproofing. I started writing. Went down there with a guitar and an amp and a little cassette recorder, and I just fucking recorded. I wrote and wrote and wrote. Bill Church was immediately on board, because he hated Ronnie for firing him from Montrose. When Denny couldn't come over anymore, I started getting his little brother, Billy Carmassi. I had this slide player, Glenn Campbell, from a band called Juicy Lucy. He'd played around Riverside before moving to England, and was fresh off the Mad Dogs and Englishmen Tour with Joe Cocker and Leon Russell.

All I knew how to do in those days was what I'd done with Montrose. That was the start of my solo years. I just knew how to be Montrose without Ronnie. I was playing guitar and singing. I was really driven to write my own songs and go out and tour until I made it. I didn't think about hits.

The first guy I called was Ted Templeman. The Warner Bros. staff producer, who made the two Montrose albums, was the only person I knew in the record business. He declined to sign me as a solo artist, but Ted did give me a couple of thousand dollars to demo out some songs. I went into Wally Heider's Studio and cut "Silver Lights," a couple of other songs that I'd written for Montrose, and a couple of the new songs. Jerry and I went down to KSAN, the San Francisco FM rock radio station, and thank God for radio stations like that in those days. They played my five-song demo—put it right on the air. I didn't even have the songs copyrighted, but it didn't matter anyway. John Carter was out there somewhere and heard it.

Carter, known universally by his last name only, was a San Francisco–based A&R man for Capitol Records, who had written "Incense and Peppermints" and "That Acapulco Gold" and a couple of other goofy songs, but I don't think he had produced anything at that point. He called and said he wanted to sign me to Capitol Records. About the same time, he signed Bob Seger. That's when Carter went through the roof at Capitol. He went on to do the Tina Turner comeback, *Private Dancer,* and then signed the Motels. Carter had some really interesting ideas. The fact that he signed me, and that I've had such a long career, shows his insight, because I never actually had any success on Capitol.

We cut the first album at the Record Plant in Sausalito. One day I saw Bill Wyman of the Rolling Stones recording stuff for his *Monkey Grip* solo album. I'm fucking starstruck. I went up and introduced myself.

"I'm from Montrose," I told him. "We did one of your songs, 'Connection.'"

"Brilliant, mate," he said and walked off.

The Plant was a crazy scene. I was coming in at ten o'clock at night and working until six or seven in the morning—the cheapest time I could get. There were so many drugs around there, it was unbelievable. One night I walked in and the guy at the front desk was doing nitrous oxide. The engineer that Carter hired was also doing nitrous while he mixed the album. He took one song called "All American" and, while I was gone, had everybody overdub. He doubled everything. He doubled the bass, doubled the drums. He put two twenty-four-tracks together. He played it for me the next day. It was out of sync and wobbled. As if the nitrous weren't enough, these guys had been on a two-day coke/weed run. I threw it out.

One day I was heading out of the studio when I heard someone buzzing at the door. The receptionist on nitrous was nowhere to be found, so I buzzed the guy at the door in. It was Van Morrison.

"Fucking drug addicts," he muttered as he walked past the reception area. I chased after him.

"Van, I'm Sammy Hagar," I said. "I'm doing a record with John Carter"—he knew Carter—"do you have any songs?"

"Like what?" he said.

"Like 'Into the Mystic,'" I said. It was my favorite Van song.

"Follow me," he said.

He picked an acoustic guitar and we went into a little tiny room. He played me "Flamingoes Fly." Giving it up—not like going through the motions. Eyes closed, singing the fuck out of it. I'm goose-bumped. This guy's my hero, my favorite songwriter at the time—him and John Lennon, they were the guys I wanted to write like.

He told me he would come back the next day and make a recording of the song for me. I was jacked out of my brain. When I told Carter, he went nuts.

When Van Morrison came back the next day, he was in a different mood. He went in, without a click track, sat down at a microphone, played acoustic guitar, and sang the song for a demo. This time, he couldn't have cared less. He knocked it off and split. But Carter gets a bright idea. We have Jimmy Hodder from Steely Dan overdub drums, Bill Church overdub the bass (he'd played with Van before), and me singing, and put together this track like it's a duet with me and Van Morrison.

We were getting ready to come out with that record and Van got wind of it. His attorneys took him off that record so fast. We had to go back in at the last minute and start from scratch on that song.

About two years later—Van's still my hero, but I never talked to him again—I went to the Mill Valley movie theater. Betsy and I buy our tickets and go stand in line. Guess who's in front of me? Van Morrison and his girlfriend. I didn't want to say anything to him, because he was shining me on. He had his back to me so hard. Within thirty seconds, he grabbed this woman. "Let's get out of here," he said, and they were gone.

The Plant always had people like that coming and going. Sly Stone had a room at the Plant they called the Pit. We worked in the room for a while—that's where the engineer did that coke and nitrous mix of "All American"—and we were told we had to move to another room because Sly was coming. One night, it was raining like a motherfucker, and I needed to get two guitars out of my car. Sly Stone pulled up in a Rolls-Royce while I'm getting my gear out of the car. He was wearing a big fur coat, floppy hat, and had an entourage around him. One of his guys goes up, hits the buzzer. As soon as the door opens, they hold it open. Sly goes in last. I run for the door. It's pouring rain and I'm carrying two guitars. "Hey, hold the door," I said.

Sly looked right at me and let the door slam shut. I was pissed. I was in a bad mood about something anyway. I went off. I started kicking the door, yelling and screaming, hitting the buzzer. Mr. Nitroushead at the front desk opened the door and I steamed past him. Sly was standing with his guys in the lobby. "You motherfucker," I said. "You could have held that door open. It's raining outside."

"People hold doors open for me, motherfucker," Sly said.

His big bodyguards swung around and pushed me against the wall and Sly walked off. I was all by myself. What a prick.

There was a girl who also worked behind the reception desk sometimes, who was always doing coke. "You want some?" she would say. She was there one night when I came staggering out of the studio after singing seven hours straight. My head was killing me. "I've got a singing headache," I told her.

She came around from behind her desk, undid my pants, and started blowing me, right there, in the lobby, about two in the morning. She wanted to take me in the Jacuzzi, but I didn't go for that. I wasn't that promiscuous then, but when a chick unzips your pants and starts going down on you, it's really hard to say no. That's the kind of place this was.

When I got my record deal from Capitol, they gave me $50,000. Before then I'd been broke—flat down to nothing and getting unemployment. The next week, I got the check for fifty grand. When I told the unemployment people about it, they said, "What about the next week? Are you getting paid?" They gave me my checks. And I've got a Porsche parked out front.

Because I was using a Capitol Records in-house producer, all the recording costs went on their budget and I got to keep the whole fifty grand. I was living good. This was the most money I'd ever had. Of course, I had to pay band members. I had to pay roadies, if I wanted to use roadies. I had to rent trucks. But the studio time and the musicians' studio pay was all covered by Capitol. They were running a tab, though I didn't realize that.

Anyway, my first album, *Nine on a Scale of Ten,* got done and came out in May 1976. I went out on tour almost immediately with Joe Cocker, Ted Nugent, lots of others. I opened for everybody. They pulled the plug on that record at 27,000 copies. It went out of print. Not because it was dying. It didn't do that great, but there was some kind of behind-the-scenes politics with Dee Anthony, who still thought I owed him money from Montrose and managed to wield considerable power in the industry, that killed it at the label.

As a result of this, I parted ways with Jerry Berg and signed for management with Ed Leffler, the man who would handle my career for the rest of his life.

I'd first met Leffler when I was auditioning for Capitol in Hollywood at the Starwood with my new band, which I was calling Sammy Wild and the Dust Cloud. A dust cloud is the beginning of a star. I was still on the Bowie kick, and I was going to be from another planet. I was going to be from Mars. I was appearing on a showcase gig that Jerry had helped put together with Back Street Crawler, a new band led by Paul Kossoff of British rockers Free, whose "All Right Now" was a favorite of mine. Leffler ended up at the showcase because he managed the red-hot British teen pop band the Sweet, and he'd come looking for a good opening act for his band when the Sweet hit the Santa Monica Civic the following week. He caught my show.

"There is no way I'm going to let that blond-haired energetic motherfucker open for my guys," he said. He hired Back Street Crawler instead, only Kossoff died the next day of a drug-induced heart attack on a plane to New York and I got the gig anyway.

At the show, I'd done everything I could. I'd run out into the audience. I'd pulled every trick I knew. I really worked that show hard and we drew a huge encore—a band without even a record deal. Backstage I overheard Leffler reaming out the vocalist for the Sweet. He was bombed, probably on drugs, and Leffler laid into him. He was worried that the band, which was huge in England and had "Fox on the Run" on the charts in this country, was too poppy for America. When we talked later, I told him I already had a manager, but I wanted to go with Leffler. I still owed Jerry Berg the $10,000 he'd paid for the showcase date in Hollywood. He maxed out his credit card to pay for it. Leffler covered the $10,000 and took over my management from then on.

Instead of fucking around, we went straight to England, to Abbey Road, and recorded my next album, *Red*. When we were getting ready to head over, Carter found a guitar player called Scotty Quick, who had a bad cocaine habit, although I had no idea. I didn't know anything about cocaine in those days, other than I'd done it a couple of times to little effect. At this time of my life, I was not into drugs at all. I didn't drink. I didn't do drugs. I didn't even drink wine. Nothing. Scotty Quick was a good guitar player, but he couldn't remember the songs. We'd rehearse them one day, everything was great. The next day, it was like a brand-new song. We were getting ready to go to England to record and he kept fucking up.

One night he came to me in my dream and I told him off, but he was vague, hard to reach, and I couldn't communicate. Next day I found out that he'd OD'd shooting coke in a Union 76 gas-station bathroom. Real quickly, we lined up a new guitarist. My drummer Scott Mathews recommended some guy he knew, named John Lewark. I took Lewark, Mathews, bassist Bill Church, and Alan Fitzgerald from Montrose on keyboards and went for six weeks to make a record in England.

We were all on a shoestring. It wasn't like I was some big star. I had the record deal, but there were all these guys that looked at me like, "I'm better than him, how come he has a record deal?" I was working my ass off. I got a record deal because I went to bed at night writing songs. I woke up in the morning writing songs. I spent every second of my waking hours trying to write songs. I really wanted to be a songwriter. Mathews and Lewark looked down at me; "I can play guitar better than him," or "I can write songs better than that." There were guys around me every now and then that would come into my world, trying to make a little money, have a job, but at the same time didn't respect me. Those guys never played with me again.

Carter had just signed Bob Seger. He sent me a demo of some song Seger wrote but didn't like, called "Night Moves." I listened to it, worked on it for a day or so, but didn't feel it. I wanted to rock. Carter thought Seger's song was a hit, but I gave it back to Carter and he made Seger do it. Carter was always trying to get me to do hits, and every time anything seemed like a hit to me, I hated it. He wanted me to do "Catch the Wind," and I loved

Donovan, so I did that, but I always wanted to be heavy metal. I wanted to rock hard. Carter was trying to get me a pop hit.

Before I went to England, I'd sold my Porsche for $5,000. I'd bought it for $5,000 and I'd sold it for $5,000. I'd heard that you could buy a Ferrari in England for about half what they would cost over here. In those days, you could go to Europe and buy European cars—Aston Martin, Ferrari, Lamborghini, Maserati, Jaguar—pay to ship them back and still double your money. I'd never even really seen a Ferrari up close. J. Geils took Ronnie and me for a ride around Boston one time in his, a 250 Lusso. That blew my mind—the way it sounded, the way it smelled, the whole thing about it. I had an infatuation. I was always a car guy.

With my $5,000, I bought a Ferrari 330GT 2+2 that belonged to land speed record holder Donald Campbell, my first Ferrari. Right-hand drive, four-seater, four headlights, bluebird blue. I bought it the first week we got there, and that pissed off Scott and those guys, too. It had four seats in it. Every day we drove to the studio in my right-hand-drive Ferrari. I was driving it all over England on days off. Betsy and Aaron had come with me to England, and I took them to Stonehenge one day. I drove to Scotland another. Betsy and I stayed out by Wembley, in a kind of apartment, so I was always driving to and from the studio. For six weeks, we went everywhere in this Ferrari. Then, right before I left, I shipped it home.

It broke down a couple times. One time, the radiator hose vibrated loose, hit the fan belt, and drilled a hole in that son of a bitch in the middle of London. I was underneath the thing in the thick of traffic, Betsy and Aaron sitting on the side of the road. The car got so hot, it vapor-locked. I fixed it with a makeshift hose and a borrowed screwdriver.

Capitol paid for everything, but I was digging a hole with them. They paid for making the records. Then they would put me out on tour opening for anybody and everybody and they had to pay for that, too, since I was only making $500 to $1,000 a night as an opening act, and it took about $1,200 a day to be on the road. So Capitol had to pay the difference, which was why, even though I was doing everything cheaply, I still wasn't making any money.

AS WE WORKED on the second album, I wrote the song "Red." Carter tightened up some of the lyrics. He was good with lyrics, so I would listen to him. He was more lyrical than me. "Crimson sin intensity" is one of Carter's lines. I thought that was clever, kind of deep. I started wearing red, painted my guitars red. I just started going red, red, red.

I realized at some point that I really love the color red. I was still into numerology from that book that I found in that trunk behind my house; between that and the dream I'd had with my father right before I heard about his death, I'd started seeking out the mystical. Somewhere it came to me that the color red was my color. That was the magical color. Red was everything.

Red is fuzzy, if you look at it. You light red with a red light, it doesn't have hard edges, like most colors. It turns into fuzz. It isn't like a defined circle. It gets deep. It looks soft. Yet it's aggressive as hell. It's blood. And it's energy. It means so many different things. Red is my color. It means everything for me. I dream in red.

I took it to numerology. R is a 9, E is a 5, D is a 4. Red's a 9. I became the red/9 guy. That was it. They both mean the same thing. They have a power. Red has a rhythm. I put red as my color, nine as my goal. I want to raise my consciousness to the nine. I changed the name of my publishing company from Big Bang to the Nine Music, wrote the song "Red," and started dressing in red. I thought this was going to represent what was going on inside of me. If I put on a pair of red pants, red shoes, red shirt, red guitar—that's Sammy Hagar. I just felt it. I believed it. No one told me to do it. It was what I wanted to do.

Years later, David Geffen told me he thought I should lose that red thing.

I got along really well with Geffen. He was a strange creature, that's for sure, but he is also as smart as they come. But, by the time I got to those guys, they didn't quite understand where I was coming from. I was deep into it by then.

Right before the January 1977 release of the *Red* album (official title: *Sammy Hagar*), I got offered the Kiss tour, very big deal, at the last minute. I was added to the bill that February at Madison Square Garden in New York so late I wasn't even advertised on the show. People started booing before I walked out. I had never played New York except with Montrose. They didn't even know me. I looked out and the whole place was dressed like Kiss. They've all got their makeup on. They were booing and flipping me off.

"Hey, what the fuck do you people think you're doing booing me?" I said. "You haven't even heard the music yet. You don't even know what the fuck you're talking about."

All dressed in red, I played the first song, "Red," entirely unknown to the crowd. I played "Bad Motor Scooter"—pulled some Montrose out of my ass. I could hear the crowd between songs: "Fuck you, fuck you, fuck you." Then I went into "Catch the Wind," the Donovan song that was the single off *Red,* and in the middle of it, they drowned me out with boos and started throwing shit onstage. I stopped.

"I'm so happy that they flew in this special audience for me from Los Angeles," I said, and the place went nuts. They charged out of their seats. They wanted to kill me.

"Fuck you," I said and dropped my pants, pulled out my dick, and smashed my fucking '61 Stratocaster to pieces onstage. What an idiot. Demolished this vintage guitar and walked offstage.

Standing in the wings, of all people, was Bill Graham, holding his face, going, "Oh, my God. Sammy." He followed me into the dressing room. "What is wrong with you? Don't ever do that! You could have won over those people."

Bill had been riding in a limo on his way to the airport when he'd heard on the radio that Sammy Hagar was opening that night for Kiss at Madison Square Garden. I often wondered if Graham was the Bill that the psychic Miss Kellerman asked me about. Graham, the promoter behind all those historic sixties concerts at the Fillmore Auditorium and the boss of the San Francisco music scene, had taken a personal interest in my case. He was working me up the bills at Winterland, his home base of operations through the seventies, and was one of the first concert producers to act like I might have a shot.

When Graham heard on the radio about my show with Kiss, he turned around and came straight to the concert—just in time to walk in and see me pulling down my pants.

As I was standing there with Bill, Paul Stanley from Kiss came in. "What happened?" he said. "That was terrible. I can't believe it."

I told everybody to shove the Kiss tour up their ass, and never did another date with the band. It was the worst experience I ever had onstage and it ruined me in New York. They didn't even know who I was. They hated me before they heard me.

I was just beginning to figure out who I was. Shortly after the *Red* album was released, I did a $1-admission concert called the Rising Star in Seattle, a radio station promotion run by some disc jockeys in the Northwest. It worked well for me. I sold out and the Northwest became one of my first big areas.

The reviewer covering the Rising Star concert in the paper

called me "The Red Rocker, Sammy Hagar." Some kid came up to me with the newspaper, and he asked me to sign it. "Will you sign it 'The Red Rocker'?" he said. I was happy just to sign an autograph. A few days later, I was walking down the street in Texas and somebody yelled out, "Hey, it's The Red Rocker." And it hit me—hey, that's me.

6

I CAN'T DRIVE FIFTY-FIVE

I found a home in Mill Valley. I had been renting this place, but when the lady who owned the house decided to sell, I knew I had to buy it. My heart and soul were already in it. Leffler arranged for Capitol to give me an early advance on my next album after *Red,* so that I could put the down payment on this architectural wonder of a house. It was on top of Mount Tam in Mill Valley, called Tamalpais Pavilion, and I could barely afford the house, but I knew I wanted to live there. A Frank Lloyd Wright protégé named Paffard Keatinge-Clay, a British-born architect, built the place for himself, got divorced, and lost it. He built this house out of cement and glass, and steel-reinforced concrete—the first prestressed concrete house in architectural history. He built another one just like it in Switzerland, and he built a bank in Pasadena that's exactly double the size. And that was it.

Glass all the way around, eight concrete columns hold everything up. The roof is sitting on top of it. It's not bolted down whatsoever. I scraped up the $60,000 down payment from Capitol and the owner carried me for the other hundred grand. I didn't know how I was going to make the payments, but I managed. I still live there today.

When it was being built, a filmmaker named John Korty lived down the street and watched the endless parade of cement trucks driving past his door. He wrote a script about a guy whose house burned down. An earthquake destroyed his next house, and then another one the termites ate and he was a termite inspector. He freaked out and built this cement house that was bulletproof. That's the story of *Crazy Quilt,* a kind of cult classic among early American independent films. That's my house in the movie.

The *Red* album had a little success. It sold about 100,000 albums. After the tour, Carter and I went straight back to England to make *Musical Chairs.* That album had a sort of Top 40 hit, "You Make Me Crazy," that was me trying to write like Van Morrison.

When it was time to hit the road in support of *Musical Chairs,* I landed the Boston tour. The group was the big new rock band of 1977. First, they went out opening for Black Sabbath, but quickly graduated to headliner. Boston hired me as the opening act for the whole tour. The first leg lasted nine months, with a break for Boston to go in and record their second album. While Tom Scholz and the rest of Boston did that, I went out and did a miniature headline tour and did pretty damn good. Small arenas, three- and four-thousand seaters in Texas and Southern California. Then I went back and did the second Boston tour for eleven months starting in fall 1978, opening every night, two and three nights in every venue in America.

There were only a few places where I could pull that off: San Francisco, San Jose, Santa Cruz, Santa Monica, San Bernardino, and San Antonio, Texas. Those were the six markets where I could go make five grand. Leffler would put me in those markets while I was out touring, opening for other people, and do quick little headline shows to keep me alive.

On that first little headline tour, I did the craziest thing. I made a live album, *All Night Long,* and that became my next release in 1978. Oddly enough, the live album sold about 250,000 records.

I was starting to break. You could see it. My record was selling with no singles, no radio airplay, no nothing. Just twenty-one months of nonstop touring.

On my next album, *Street Machine,* I parted company with Carter and decided to produce it myself. For years, Carter had been giving me these dumb songs, always trying to get me a Top 40 hit, trying to get me to do covers like "Dock of the Bay." When I'd finally bent over backward and done "Dock of the Bay," covering Otis Redding, for God's sakes, with guitarist Steve Cropper, who wrote the damn song, it didn't even work. I'd brought the guys from Boston over to Wally Heider's Studio in San Francisco after a May 1979 Day on the Green concert we played across the bridge in Oakland before fifty-five thousand fans, and they sang background vocals. That was supposed to be a shot at a Top 40 thing, but even KFRC in my hometown would not touch it. I was already headlining concerts in the Bay Area for promoter Bill Graham, but they wouldn't play my records on the radio. I'd given Capitol "I've Done Everything for You" on my live album, a song that two years later was a Top 10 hit for Rick Springfield, but they hadn't been able to get one radio station for me. To Top 40 radio, I was a heavy-metal guy.

By the time I went into the studio to record *Street Machine,* I was over trying to get a Top 40 hit. That record sold about 350,000 when it was released in September 1979. After years of opening concerts for everybody and their uncle, people started to think I could be a headliner. Louis Messina of Pace Concerts in Texas packaged me with Pat Travers, who was on the charts with the hit "Boom Boom (Out Go the Lights)," and the Scorpions, the German hard-rock band who were just starting out in this country. We sold out everywhere. It was unbelievable. We were doing ten, twelve thousand seats. It was a low ticket price and it was a package deal, but it was the first time I did a headline tour.

But Capitol Records still didn't get me. I had just done a headline tour, selling out arenas, and they couldn't get me past 350,000 records. My business had quadrupled, not just the box office but T-shirts, everything. I was becoming a genuine rock star onstage. In England, I was on the cover of *Melody Maker* and *New Musical Express*. They took a picture of me in shorts, high socks, tennis shoes, and a tank top with my Trans Am and my Explorer. I looked fucking mean. One headline said, "Van Halen, Look in Your Rearview Mirror." We sold out fourteen theater shows in England before we even left this country.

My last album for Capitol, *Danger Zone,* released in June 1980, simply reinforced that Capitol did not know what to do with me. It sold another 350,000 copies, even though I was packing venues all over America. I sold out the Oakland Coliseum Stadium that Fourth of July. Originally, Tom Scholz, the genius guitarist behind Boston, was going to produce. He came out and did preproduction, but his record company decided he should be working on another Boston album, not somebody else's record. They were going to sue him, so he left. The next day, I hired somebody I knew, named Geoff Workman, at the last minute, because we were already kind of in the studio, ready to go. Workman, an engineer who worked with Queen's producer, had just finished recording an album with Journey and their new lead vocalist, Steve Perry, and I liked what I heard. Scholz was upset that I replaced him so quickly, but I told him I didn't have money to burn. I needed to get going and get back on the road. That's when I really broke wide-open. Touring in support of *Danger Zone,* I saw it was really starting to happen for me. I saw it with my own eyes. At every concert, people were singing my songs. They knew who I was. I could see it coming every night. From the time I began playing music, I'd put my nose to the grindstone, head down, rolled up my sleeves, and went forward. I'd never made it and I'd never had any money. When I came back, my accountant told me I had $300,000 in the bank. "What do you want to do?" she said.

I'd made an album a year for the five years at Capitol, while I was touring constantly. I would come off the road and go in the studio. If I wasn't touring, I was making a record. The label paid tour support, but because my records weren't selling that well and I was constantly on the road, I wasn't earning out the expenses. I had done well in England and other European countries, but Capitol never paid me a single royalty. In fact, they told me I owed them $175,000. I had a really bad record deal. I was getting about twenty cents a record. I was spending more money on tour than I was earning. I decided to sue Capitol.

John Kalodner, the big-cheese A&R man at Geffen Records, wanted to sign me to the label David Geffen had just started. At that point, he had only signed John Lennon and Donna Summer. They offered me a million-dollar deal. I was getting fifty grand a record from Capitol and owed them money, but the money Geffen gave me paid for the lawsuit. One day, we walked into court at Marin Civic Center and the judge told Capitol, "I think you folks have made enough money off this young fella." He let me out of the deal. I walked away a free man.

Geffen broke me on the charts. I finally had hit records that matched my box office on the road. Capitol never managed that. Capitol didn't market me. They didn't give a crap. It was especially sweet when I signed with Geffen and shoved it up Capitol's ass with my first gold and my first platinum album. That was the beginning of a sixteen-year streak of million-seller albums.

Kalodner was the greatest A&R guy. He got Jimmy Peterik of Survivor to cowrite a song with me, "Heavy Metal," and sold it to the movie even before my album came out. Jonathan Cain of

Journey and I cowrote a song, too. Kalodner tried to put me to-gether with different writers, but I didn't like writing with other people. I was not that confident to be sitting around a guy I didn't really know and show him my ideas. I would start tightening up. I didn't feel like I could express myself well enough. Besides, I take everything personally. But I wrote twenty-eight songs for the album. Kalodner suggested Keith Olsen as producer and I liked that idea. Olsen produced Fleetwood Mac and Pat Benatar. We made a great record, *Standing Hampton,* no question about it, with an instant Top 40 hit, "I'll Fall in Love Again," when it was released in January 1982. Went out on tour, headlining arenas, double nights in a lot of places, and I became rich and famous right then and there.

After Geffen signed me, everything changed. The turn of fate that happened in my life was unbelievable. I had plenty of money from my record deal with Geffen, and Kalodner didn't want me to even think about going on the road. For the first time in my adult life, I was home for almost a year. Since I started in the business, I'd never been home. Betsy was as happy as a lark. I built a swim-ming pool for my house. I grabbed my old pal David Lauser, who was with me in the Justice Brothers, and put a band together. I had money. I had a band. I had a crew.

For years, I'd been working my ass off. Family life was some-thing that just sort of happened to me. I'd barely noticed. I got married and had a baby while I was struggling with the band. When my son was a child, I went out on the road. I couldn't always afford it, but even after I could afford it, taking Betsy and a small child was never easy. When Aaron was older, we put him in a boarding school, North Country School in Lake Placid, New York, and Betsy started going on tour with me. But she couldn't stand touring. She hated flying. She didn't like hotel rooms and living out of suitcases. My marriage was always a struggle. If she wasn't on the road, every night calling home meant arguments. I was messing around a lot, as much as I could. I still wasn't doing drugs or falling down drunk, but I started living the life a little bit. I tried not to get involved in anything serious. That way I could at least convince myself that I wasn't really doing any damage.

Almost across the street from Aaron's school, on the highway, there was this log cabin that was for sale and wasn't expensive. It was a log-cabin kit with a big loft, five acres in the back. I bought it and we tried to spend as much time as possible there, but I was always on tour. We'd go back there for Thanksgiving. That was the law. They didn't allow students to go home for Thanksgiving, because they had this big-deal Thanksgiving feast that the kids prepared. We went back there for Thanksgiving. That was about it.

SUCCESS REALLY MOTIVATED me. Ed Leffler was amazed. "You're different than anyone I've ever met in this business," he told me. "Fame and fortune inspire you. You get better. I've never known anyone in my entire life like that in the music industry. The more success you have, the better you get. You jump on that stage now. You're so much better than you were when you were hungry."

When I was hungry, I lacked confidence. I was afraid to let my heart and soul out. I was hiding. I was faking it. It seeped through. You could hear it in my voice. My actions were not true and honest, so they didn't connect. I was bluffing, acting the part. It took fame and fortune for me to become myself. That gave me the confidence I needed to bring out what I really have to offer, whatever it is. I started to get more real.

I took what was left of that big money advance from Geffen and started buying property in Fontana. I didn't think I would have that much longevity as a rock star, and I never wanted to be poor again. My mom instilled that in me—you've got to have something to fall back on. I started building apartment buildings in Fontana. I went to my brother-in-law, James, who was an electrical contractor, and had gotten a contractor's license. My nephew also became an electrician, and one of their friends became the plumber. I made them all partners. We built nine apartment buildings. I bought the old houses we rented when I was growing up. That was my first entrepreneurial effort and we did really well.

Shortly after we'd started building the apartments, the fire department came to my brother-in-law and said he needed to put a fire hydrant in front of every apartment building. He told the fire department that his plumber could put fire sprinklers in the building that would be more effective for about the same price. The insurance companies went along, because sprinklers put out fires before fire departments could even get there, but the fire department needed some convincing. We staged a demonstration for them. We bought one of my old houses, sprinkled it, and then lit a fire in a trash can. We waited for the neighbors to call the fire department, which was parked, waiting, right down the street, and, by the time they got there, the sprinklers put everything out. The house was still totally cool. Fire sprinkling is amazing. It really saves lives. The city passed an ordinance and gave us some money. Before long, we had 180 employees and ran the second-largest fire sprinkler company in America, Fire Chief Inc.

The next thing I did was the travel agency. I started a travel agency because I was traveling so much for tours that I was paying my travel agent a small fortune. I decided to start my own, Steady State Travel in Mill Valley, hired the two ladies that used to work for the old travel agency, and gave them a piece of the action. It didn't make a lot of money, but it also didn't cost me anything when I went on tour.

I started Red Rocker Clothing, which was a disaster, because

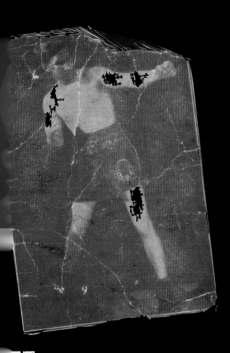

My dad, Bobby Hagar.

The Hagars before I was born, 1945.

Bob Hope and Bobby Hagar (holding me as a baby) in Palm Springs, 1948.

Baby Sam.

The Hagar kids (*from left to right*): Bobby Jr., my sister Bobbi with me as a baby, and Velma (with cat), 1947.

My fourth grade photo.

With my stepdad, Mike, at my mom's place in Cucamonga.

My high school graduation photo.

The Fabulous Castilles in my Anastasia Street backyard in Fontana.

The Fabulo[us]
Castilles
Sam, Ed Mat[t?]
& Jerry Mart
Fontana C
Late 60's

At the reception for my wedding to Betsy with my two nieces looking on. The reception was at my mother's house.

Betsy and me with our first son, Aaron, in 1970.

The Justice Brothers in 1970.

Montrose's first publicity photo. *(Photograph courtesy of Getty Images)*

Montrose at Wembley Stadium, London, opening for the Who in 1974.

At the Record Plant in Sausalito where I recorded my first album.

Behind the wheel of my first Ferrari 330GT 2+2 outside Abbey Road Studios in London. Aaron is with his stuffed bear, Theodore.

The Red Rocker opening for Boston, 1977.

With my Trans Am in Mill Valley, 1979.

Onstage and working hard.

No singles, no radio airplay.
All it took was twenty-one
months of nonstop touring.

At the top of the Marin
Headlands with my mountain
bike.

Day on the Green concert, Oakland Coliseum, July 4, 1980.

Governors' Camp, Kenya, 1983.

Egypt with Betsy and Aaron, 1983.

Hagar, Schon, Aaronson, and Shrieve (HSAS) at the Warfield Theater, 1983.

Full '80s—*VOA time, 1984.*

Bringing Andrew out for the encore at the Cow Palace, San Francisco, which was toward the end of Van Halen's 5150 Tour.

Aaron and Andrew in the back of the limo on the way to a Van

A portrait by Annie Leibovitz, 1986.

With Eddie on the OU812 Tour, 1988.

the rag trade is the craziest business in the world. I had this great idea to make these upscale flannel shorts. I bought the flannel from Ralph Lauren. He had this line of flannel shirts and stuff, and it was the baddest flannel. I lost probably $300,000, because I got a huge order I couldn't fill from JC Penney's. I was late. I ended up with $65,000 worth of these flannel shorts in my warehouse, because they wouldn't take them. Some of them didn't have buttons, I was trying to rush them out so fast. The next year, everybody had flannel shorts. Very tough business. You come up with an idea and the next year everybody flat rips you off and you have to come up with something fresh. I bowed out.

While I lost some money on the clothes, I ended up starting something else that made money: bike stores. It was Bucky who got me into the bikes. Bucky, my old pal who I used to help steal albums from the ABC Store and who turned me on to *Fresh Cream,* was living in an apartment on B Street in San Rafael with his wife, Joelle, and their son, Benny. He'd married her in Rochester, but she ran away with some other guy and they split up for a while. Bucky took her back after he moved to California, which is when they had their kid. Bucky was always around. I took him to England with me as my roadie and truck driver when we did *Red.* Ed Leffler loved him, but he was tough on the other guys. He was hard-core and always looking out for me. Plus he was into drugs and drinking and could be an asshole. We had to cool him out from time to time.

Eventually, Bucky took a job at this bike shop, the Corte Madera Cyclery, an old-time Schwinn dealership. This was right around the time that two guys named Steve Potts and Gary Fisher were inventing the mountain bike in Marin County. They took a fat-tire, cruiser bike and put gears on it from a ten-speed. They rode these bikes up and down Mount Tam. One day, Bucky took me in the back of Corte Madera Cyclery and made me a mountain bike. Right away, Bucky and I were riding our bikes everywhere. He biked to work every day, rain or shine. Between his biking and mine, I saw the mountain bike business coming and it really appealed to me. He told me I could buy the store for around $75,000, pretty cheap. All I had to do was buy the inventory, and it was a small store. I bought the store and he started making mountain bikes. We were the mountain bike kings. All these guys were bringing their cruisers to Bucky and he converted them to mountain bikes. At the store, Bucky couldn't put mountain bikes together fast enough. We had to hire mechanics.

Seeing the success of that store, I had an idea to open an even bigger bike store, a superstore that would carry bike clothing and accessories. I built the Sausalito Cyclery. We were the number-one independent bike store in California, one of the top ten in the country. We were doing $4 million a year in sales out of that place, with a million dollars in inventory on the floor. I bought the top shit. You couldn't even get into the store half the time. We were blasting twenty to thirty high-end bikes a day out of there. When the first commercial models from Specialized came out, mountain bikes started taking off. We bought them. We started seeing the trend. People started buying these things. They were trading in their road bikes. Pretty soon we couldn't even take trade-ins, because nobody wanted road bikes anymore.

We put mannequins in the store with the bike clothing and we got a review in a bike magazine saying we were the only bike store in America that displays clothing on mannequins. I made my own mountain bike—the Red Rocker. I landed the cover of *Mountain Bike Magazine* with this thing. I had two lights on it and I was the first bike builder to use black components. Before the Red Rocker, everything was chrome. You had the bike, whatever color. You had black tires, sometimes white walls. Everything else was chrome. I wanted everything red and black, no chrome. It took me a year and a half. Gary Fisher made my frames. There were different gear people in Japan. They made me enough parts for one hundred Red Rockers, two water bottles, two Maglites, and all-black components—rims, spokes, bolts. It was a really bad-ass machine. We sold out instantly, bang, gone. We had ten thousand back orders from around the country.

We went back to Japan, where our suppliers told us the most they would make was three hundred this year. Meanwhile, here comes Specialized with their red-and-black Rockhopper—Rockhopper? Red Rocker? Pure coincidence, I'm sure—and they stepped on me. They had components for fifty thousand bikes. I got out of the business, but I was pissed. I had been totally ready to take over the mountain bike world.

The Sausalito store was a gold mine, but Bucky wasn't running the place. He couldn't. He'd show up late and yell "fuck you" at someone and walk out. Everybody loved him, but you couldn't put him in charge. Instead he worked the floor. After months of this, I finally had to sell Corte Madera, because Sausalito killed it. Everyone came to Sausalito, because it was built right on the bike path.

With all these different businesses going on outside of my music, I was making some money, and I began buying things. I bought a couple of other houses beyond the one in Marin County. I started getting into Ferraris. I started developing a taste for fine wines. One night, when I was in Montrose, I'd tasted a 1945 Latour and a 1927 Martinez port on the same night and I started to build a collection of fine wines. I made concert promoters provide me with certain vintage bottles backstage as part of my contractual requirements and take them home unopened. Bill Graham was hip to my chisel. He had all five bottles in my dressing room opened, so I couldn't take them home, and, later, gave me a recorking machine as a gift. I just started living the life. Betsy was able to spend a lot of money, too. She was spending money to keep herself happy. She'd go shopping and refurnish the house. I'd come home and go, "What?" but, since I was doing okay financially, I didn't really care.

Things still weren't great between us though, and around then, I finally had an affair. I'd been screwing around on the road here and there for years, but this was different. This was a real affair, where I fell in love with another person. She was in the record business. I'd met her when I was recording my first album with Geffen in 1981. She represented a music publisher and maybe had a song for me. She was so independent. She lived by herself, owned her own house, drove a new car, worked hard at a good job—the opposite of Betsy. I fell in love with her and I began a long-running affair. I'd fly her out on tour. Betsy would leave, and she would come in. I used to fake trips to Los Angeles to see her. I'd fly down for the day. She'd pick me up at the airport. We'd go to her house and have insane sex. She was so liberated—I loved that about her. It was like, my God, this woman can take care of me.

After the affair had been going on for two years, I was ready to leave Betsy, but then I decided that, first, we needed to take a family vacation, this big trip to Africa. Betsy, Aaron, and I went to Italy, Sardinia, Egypt, and Kenya, where we spent six weeks on safari. We were gone the whole summer of 1983. I was looking to figure out what I was going to do with my girlfriend. I needed to figure myself out. I was planning on leaving Betsy. I was in love.

We were in Sardinia. Aaron was out at the pool and Betsy and I had a quick daytime throw-down. It was a beautiful day and everything was right. I knew immediately she was pregnant. That had happened the first time with Aaron, too. That time, we did it while we were listening to Procol Harum's *Salty Dog* album on a tiny record player in a hotel room, and afterward I just knew. In Sardinia, it wasn't like it was amazing sex or something—it was actually a kind of a quickie deal—but you could tell something happened.

Sure enough, we get to Africa and Betsy's kind of sick all the time in the morning. We arrived at the Mount Kenya Safari Club just in time to see Robert De Niro leaving. He was with this little kid and a white-haired guide in a Land Rover, pulling out as we pulled up. It was British-style Old Colonial. I hated the place. After five o'clock, men were required to wear a coat and tie. Women had to be in evening gowns. Children weren't even allowed out of their rooms after five o'clock. Betsy was sick. She couldn't leave the room anyway. They treated you real well and the place was gorgeous, but it was stupid fancy. It was a bird sanctuary. They had these black guys with white gloves in tuxedoes going around with little brooms and buckets, sweeping up bird poop. But best coffee I ever had in my life? Mount Kenya Safari Club, no question.

Anyway, we went all over Kenya and Tanzania on these safaris, like the Governors' Camp, which is camping, but very elegant. De Niro was there. He was on the same safari I was, either coming or going. A couple times we'd see each other in the bar, had a couple of words. "Hi, I'm a big fan, yeah." He didn't know who I was, but he knew I was somebody. Long-hair fucking hippie-looking dude in *this* place.

Going home, we flew from Kenya to London, fourteen hours, and changed to the Concorde. I was really splurging. First class all the way. Flew the Concorde to New York, changed planes for Albany, where I picked up a rent-a-car and started driving to our log cabin in Lake Placid, where Aaron went to school. We were taking him back to North Country School to start in September.

It was two o'clock in the morning when the cop pulled me over. We'd been traveling for twenty-four hours. I was burnt. While I was out of the country, they had changed the speed limits. The cop starts writing me the ticket. "Officer," I said, "I was only going sixty-two."

"Around here," he said, "we give tickets for sixty-two."

He was parked behind some trees on a four-lane highway, nobody on the road, the middle of the night. I looked at Betsy. "I can't drive fifty-five," I said.

As soon as I heard myself say it, I went, "Whoa!" Grabbed some paper and a pen. I started writing the lyrics. As he's writing the ticket, I'm writing the lyrics. The cop came back. He handed me the ticket, and I said, "Thank you, sir."

I went straight to my house in Lake Placid, a three-hour drive from Albany. By the time we got there, it was about five o'clock in the morning. I had a guitar and an amp in my basement. I went downstairs, picked up my guitar, turned on the little tape recorder, and wrote the damn song, right there on the spot.

The whole time we were in Africa, I'd been writing songs, because I knew I was coming back to do this thing with Neal Schon of Journey, my first idea for a super-group, HSAS—Hagar, Schon, Aaronson, Shrieve. Neal and I had it all planned. When I got back and found out Betsy was pregnant, I kind of decided to end my affair or, at least, started slowing down. I had a pregnant wife on my hands and I thought this wasn't the time to leave anybody. It ripped me up, because I really was in love with this woman, too. But the day Betsy went into the hospital to have Andrew, on June 4, 1984, I called the girl from the music publisher from the hospital and cut it off.

"I've got a new baby boy," I told her. "I'll never see you again." I never did.

When a baby's born, it is a miracle. You can read the Bible or other books, and you hear about miracles. You want something to affect you and change your life like that. You want to see Jesus walk on water. You want to see someone heal, take a cripple and make him walk. You want to see those things. We all want that. When you see a baby born, you see that.

It shipped me right into shape. I had the whole world in my

hands and I watched this baby being born. I was ready to give up anything for that, for my kids and my wife, so that we could continue to be that family together. There's just something about seeing a child being born. Creation. Isn't that as close to God as you're ever going to get?

When I turned my focus back to HSAS, I had all these songs like "Giza" and "Valley of the Kings" that I had written in Africa and Egypt, these vibey kind of lyrics, and I had "I Can't Drive 55," because I'd written it on the way home. But I didn't give it to the band. I didn't even tell Neal. I just kept it in my pocket. I wrote the whole song anyway, and Neal and I were supposed to be cowriting everything on the new project. I don't think I was being cheesy or cagey, just that was my song and I saved it for myself.

Good thinking, it turned out. The HSAS thing didn't really work. Neal and I had wanted to do something together, but I don't know why we got bassist Kenny Aronson and drummer Michael Shrieve. They were good and everything, but it was more a matter of who was available. Aaronson had played with Billy Squier and Neal knew Shrieve from when they played in Santana together. Shrieve, who's a great rhythmical guy, wasn't a rock drummer at all, and we were a rock band. But he made the band kind of cool and fusion-y.

We did twelve shows in November 1983 around the Bay Area, all sold out for a band nobody ever heard before, and gave all the money to arts and music programs in public schools. We cut the album live, which I thought was sort of adventurous, but it never sold more than 150,000, even though I was selling more than a million as a solo artist and Journey was selling more than a million with Neal. It never caught on. "A Whiter Shade of Pale," which was the single, didn't really work, never hit. We played an MTV show called *The Concert*. Neal and I went to New York and did press for days. But it just never took off. It might have been better if we'd gone in the recording studio, made the record, and then done the shows, but the way we did it was unique. We never toured again.

I went right back to record "I Can't Drive 55," which ended up on my album *VOA*. That album was produced by Ted Templeman, Montrose's old record producer who'd fronted me the budget for my first solo demos, and I recorded it at Fantasy Studios in Berkeley, this large orchestral room where Journey just finished recording their album *Escape*. I did all the demos at my little studio in my house in Mill Valley. David Lauser came over and laid down drumbeats, and the two of us would spend ten or twelve hours every day in my basement, working up a bunch of ideas.

VOA with "I Can't Drive 55" really took off when it was released in August 1984 and made my business ridiculously big. "I Can't Drive 55" was not my biggest hit, by any measure, but it means more than any song I've ever written. At the time, "55" only went to number twenty-six on the charts. It wasn't even a Top 10 hit, but it was the one that really sold the records and kicked my concert business into the stratosphere.

The tour for *VOA* was my most successful. I sold out arenas everywhere, two, three, or four nights some places, one of the top-five grossing tours in 1984—right up there with Van Halen, who broke at the same time with "Jump" and all that. I remember getting an award in Portland, Oregon. I sold out two nights and got the Show of the Year. Van Halen was runner-up. We were neck-and-neck on the road. They had a much bigger record. My album was 1.6 million, but they ended up selling 10 million records.

My records were never up to my box office. There would be a guy like Greg Kihn or Tommy Tutone, who had his moment where he sold as many records as I ever did, maybe even more. But they could never sell out arenas, and I could go out and do multiple arenas. I was a live performer who came up as an opening act. My albums were just ways to get me back out on tour, until I met John Kalodner and Geffen, who got my records going.

Kalodner and Geffen also got me doing music videos, which pushed my VOA Tour even more. The video to "I Can't Drive 55" was huge on MTV, doubled or tripled my box-office business, and did for me instantly what radio never did. It made me a star. The all-music cable channel started in 1981, but it took a few years for the idea to catch on with local cable companies and the public. Once MTV did catch on, it had incredible power, and making videos was almost as important as making records. "Three Lock Box" had been my first video in 1982, and I'd noticed it changed me from being relatively anonymous to anyone but my fans to someone that old ladies recognized walking through airports. But the "I Can't Drive 55" video took everything to a new level.

Kalodner put me together with director Gil Bettman, who had done great car scenes as director of the TV series *Knight Rider*. It was like shooting a movie—quarter-million-dollar budget, four-day shoot, twelve-hour days, at different locations all over Southern California. We rented a wing of an old hospital in Los Angeles and built a jail, put up bars, and a courthouse, where the judge sat and the old lady hit me with the umbrella. We used real California Highway Patrol officers in one shot, where they wrestled me down on the hood of the car. I got to drive my Boxer 512 flat-out. We went out in the desert by Palmdale, where I could go 170 miles an hour. Gil dug holes in the ground and put cameras in them to drive over.

That song changed my relationship with the California Highway Patrol. At that point in my life, I'd had thirty-six tickets. My license taken away three times. I was paying $125,000 a year for car insurance, because I had all these hot cars. I'd been to traffic school. I had hired attorneys. I erased as much as I possibly could, legally and financially, and I was still in bad shape. "I Can't Drive

55" changed everything. Since I wrote that song, I've maybe had two citations. I've been pulled over at least forty times, stopped and let go.

Some of the stories are classic. I was driving my Ferrari one night from San Francisco to Malibu with Betsy—this was later, during my first year with Van Halen—and was rolling between 150 and 160 all the way down Highway 101. As I approached Santa Barbara in the Ojai area, where there are all these speed traps, I decided to cut my speed. I had been checking my rearview mirror the whole way and couldn't see anything behind me.

Up ahead is a roadblock, two California Highway Patrol cars. I didn't think there was any way this was for me when I pulled over. About the same time, a helicopter lands, and three other Highway Patrol cars pull up. They had been chasing me for a while. I just didn't know it. Those little five-liter Mustangs are good for 140 miles an hour, max. I was blowing them off so bad I couldn't see them in the rearview mirror.

The cop got out of his car, shaking, with his gun in his hand. "Get out of the car," he said.

I got out. "You better have a good excuse," he said.

"Sir, I do not have a good excuse," I said. "I was just having a good time. I have a fast car. I've been to driving schools and taken racing car courses. I didn't think I was in danger. I was not reckless-driving."

He backed off and holstered his gun. He walked off, took off his hat and wiped his brow, replaced his hat and came back. "Okay, how fast were you going?" he asked.

"I was really going fast," I said. "Probably around 150."

He threw his hands in the air, and marched back to his car to the other guys to confer. Everybody was worked up. "My life was in danger," one of the other cops kept yelling at Betsy. Finally the cop with the gun sent everybody else away. They took off.

The guy took his hat off again and pulled me over to the hood of my car and leaned against it. He started telling me his troubles. He had looked at the license and knew who I was.

"You know, I've got kids and this is a really stressful job," he said. "And here you are, a rock star, your life's in your hands. You've got anything you want. I was chasing you down the road thinking, 'I want to kill this guy when I pull him over.' And then you sit there so calm. You tell me how fast you were going. You didn't lie to me."

He sat there forever, singing the blues to me about his life. At the end, he stood up, put out his hand, and said, "Nice to meet you." The wackiest pull-over ever.

Just recently, I was blasting down 101 in Marin County, going to rehearse with my new band, Chickenfoot, driving my Boxer 512 from the "I Can't Drive 55" video, doing about ninety, when I flew past a cop under the freeway. Sure enough, he pulls me over. I came to a stop on the side of the freeway and rolled down my window. He came up and squatted down, his dark visor down on his helmet. He held up his radar gun, pointed it at my face, and it was flashing "55-55-55-55." He pulled off his helmet and he has the biggest shit-eating grin on his face.

"I've been waiting for you, man," he said. "All the fellows told me you're around here and they see you all the time. I'm the biggest fan you ever had. I just got out here from the South. I've been on the force for two months now, and I'm telling you, man, I've been looking for you. And I got you!" You write a song like that, and no telling what happens. I wish I was smart enough to say I'd done it on purpose.

Even though I made a lot of money on that tour, I came home from it ready to take a year off and figure things out. Betsy was on my ass all the time; she wanted me off the road and we'd just had Andrew. I told her we would buy a house out in the country.

I went and looked at this place in Nicasio as well as a property in Big Sur. We were really going to go remote, try to get off the grid. I had been reading this book *The Coming Hard Times* by this guy who said the banks were going to collapse, everything was going to fall apart, the bottom's going to come out of society, and gold was the only thing that would be valued. Paper money would be worth nothing. I really believed this shit.

I was looking for a cabin in the woods with a hundred acres. I was stocking up food. I got guns. I had all my ammunition, not to kill people, but to eat. I learned how to kill animals. I went hunting all the time. I learned how to kill a deer and skin it, how to cure it and eat it. I was going to go Ted Nugent on everybody. I was going to hunt and fish. I was going to put a racetrack on the spread and have all my cars.

The more I thought about things, the more I decided to stop the grind—album/tour, album/tour, album/tour—for my wife and family. I was burned out. I'd been on the road for more than ten years and, before that, I'd had a hard life—work, work, work. Now I had everything going. I had my bike stores, my travel agency, my fire sprinkler business. I had my apartment buildings. My business manager had done a great job for me. I was set. I didn't even need any royalties. I was at a peak, and I felt like the fucking king. I could do whatever I wanted. I had $3 million in the bank, and I could see daylight. I would do one more record and one more tour, put away another $2 million, and retire altogether. Between all the other businesses and the money I already had, it was not necessarily a lot of income, but certainly more than enough to live on.

I was ready to give up everything. Betsy was pressuring me. I was seeing things her way, and then Eddie Van Halen called.

7

5150

I came off the tour, I was fried crisp. I cut off all my hair. I canceled the last four dates on the tour after I hurt my foot—twisted my ankle in Connecticut and couldn't walk on it. I tried to do one show like that, gave up, and went home. We had done ninety shows that year.

My Ferrari 512—the car from the "I Can't Drive 55" video—was sitting in Claudio Zampolli's shop in Van Nuys, where I'd bought it. He was an Italian mechanic, who doubled as a salesman (he's actually the guy I'm talking to at the beginning of the "I Can't Drive 55" video). He didn't run a dealership or anything, but he'd buy a car for you. He used to work for Ferrari as a test driver. It took nine months to get the 512 after I ordered it. They made, like, twelve that year. Anyway, after I'd had the car for a little while, it needed a tune-up, which, on these special cars, is a very big, expensive operation. It's really a race car, and a tune-up can cost as much as an ordinary new car. I went home without picking up my car.

Eddie Van Halen drove a Lamborghini, a Countach, and Claudio worked on his cars, too. Eddie saw my car at Claudio's and asked him about it.

"Hey man, nice car," he said. "Whose car is that?"

"Sammy Hagar," said Claudio. "You should call him and get him in the band."

Everybody knew that vocalist David Lee Roth had left Van Halen a couple of months earlier. He quit the band almost as soon as his little solo single, "Just a Gigolo," started to do something. It was too soon to say they were floundering, but their predicament was public knowledge.

"You got his number?" Eddie said.

At my house, the phone rang. It was Eddie Van Halen. "Hey man, what are you doing?" he said.

"I just came off tour and I'm just kind of nursing my foot," I said.

"Would you like to get together, come down and jam," he said, "and maybe join Van Halen?"

"Not really," I said. "I'm burnt. I'd love to meet you, but . . ."

I'd only met Eddie briefly a couple of times. We'd done a couple of big festivals together, and he'd come to my dressing room. "I'm such a big Montrose fan," he had said. "What a nice guy," I thought, "so humble and sweet." When you shook his hand, he was always holding yours with both his hands and adding in a little bow.

"How about tomorrow?" he said.

"No, man, I can't," I said. "I'm burnt. A couple of days, at least. Let me call you back."

We exchanged numbers and hung up. I started thinking. Maybe if they're broken up, I could get Eddie in my band. I could use a gunslinger like him. Or maybe I'll just write some songs with him and get him to play on my next record. I was a big fan.

But I hated Dave. The guy rubbed me wrong. I'm sure I rub all kinds of people wrong, so it's not like I'm putting him down. The guy was a great front man, great attitude in rock, and had an image from hell, but I just couldn't stand the guy. He was the opposite of what I believed in and what I am. First of all, the guy's not a great singer and he acts like he's the coolest, hottest guy in the world, when to me, he looks gay. The guy was never believable to me.

The call didn't come as a complete surprise. Ted Templeman had been the one to tell me that Roth had split a few months earlier, and at the time, I'd told Betsy, "They're going to call me, you watch." Who else were they going to get? There was Ozzy Osbourne, Ronnie James Dio, and me. When Eddie did call, I was sitting there with goose bumps on my arm. I went down to see him a couple of days later.

I walked into their place in Studio City. Alex Van Halen took one look at my short hair and started laughing. "You look like somebody put a doughnut on your head and cut it off," he said. I had the sides shaved and left just a little bit on top. I was taking a year off. Alex was drunk on his ass. He was drinking a case of tall malt liquor cans a day. He pounded them, too. He could shotgun like nobody. He always wanted to have contests. He would pass out a couple of times a day, wake up and shotgun two or three beers, crack one more, and walk out of the room. Eddie drank all day, too. They both woke up, grabbed a beer, lit a cigarette, and that was the way they started their day. Midday, around four o'clock, they would take a nap. They were both big nap-heads.

Eddie lived in a very humble house with his wife, Valerie Bertinelli, the actress. It was actually Valerie's house—Eddie just moved in with her. She also had another place, sort of a beach house in Malibu, and they split their time between her two homes. The main one though was up in the hills off Coldwater Canyon. It was just an ordinary two-bedroom house with a garage that he'd converted into a studio. They called the studio "5150," after the police code for picking up a crazy person. It was not a rock-star home, and the studio was a dump. They were recording through a homemade board that could have come out of a Cracker Jack box, built by engineer Donn Landee. Landee could make the board sound brilliant, but he was a genius and knew how to work it. To anybody else, it was like model airplane gear.

The studio was filthy. Beer cans everywhere, ashtrays full of cigarettes. Donn Landee had to blow away the cigarette ashes just to plug something into the board. The place smelled like the worst bar on the planet. I don't think it had ever been cleaned. Eddie's guitars were lying on the floor. Nothing on racks, nothing in cases, just on the floor, on chairs, leaning against amps, against the wall, a pile of them in the corner. It was beautiful, but I'd never seen anything like it.

Eddie walked in, wearing a pair of those shades with louvers in them. He'd been up all night, drinking, trying to write some music. I didn't know these guys. I didn't know what their routine was. But they were beat up. Eddie was wearing a pair of wrinkled pants. When I went into their house later that day, I saw why. He and Valerie were living out of their suitcases. They had been off the road for a few months, but they didn't have their stuff hanging in their closets. It was sitting in their suitcases on the floor. There were piles of stuff everywhere. It was weird. They could afford maids, but they didn't have them. They were kids. If you really look at it, they had been out on the road for five years and had only recently come home.

Eddie never bothered to unpack. He was always pulling clothes out, finding something halfway clean but wrinkled. I found all this kind of humorous, like, "Far out, these guys really don't care." I thought that was pretty cool. I came from a different world, Betsy's world. My clothes were pressed. My socks were ironed, folded, and put in the closet. I was actually wearing a suit—Armani linen jacket and slacks, T-shirt, and tennis shoes, kind of *Miami Vice*. I ate in good restaurants and drank fine wine. Eddie would throw a hot dog and bun out of the freezer in the microwave, nuke it, plow it into his mouth, and chug it down with a beer. There were old pizza boxes lying around. The refrigerator was full of frozen burritos.

Al was the crazy one. He was obnoxious, drunk, making comments, laughing about stupid things, smoking cigarettes. "Here," he would say, "shotgun this beer." I don't drink beer.

When I got there, they'd been up all night writing. They had what became "Summer Nights" and what became "Good Enough." Eddie, Al, and the bassist, Mike Anthony, had stayed up jamming. I arrived at about noon, and they still hadn't been to bed yet. And they were ripped. They had been drinking the whole time. I went down to check out Eddie. In my head, there was no way I was joining Van Halen.

We started playing, and the engineer Donn Landee recorded everything we did. I made up the first line on the spot—"Summer nights and my radio." It just popped into my head the first time I heard that riff. The rest of the song I scatted my way through. I did the same thing with "Good Enough"—I really had my scat together. Eddie couldn't believe it. Dave apparently didn't have good rhythm and wasn't a great singer, didn't have any range. I was singing Eddie's guitar licks with him. After five hours, they were freaking out.

"We've got a band," they kept saying.

"I don't know," I said to them. "It sounds great, but let's talk about it. Maybe I'll come back next week or something."

They wanted me to stay, but I went home and took a cassette. We jammed a blues song and we had the other stuff that we worked up. After dinner, I put the tape on my stereo. I got the goose bumps all over my body. I heard it. I realized it was Cream all over again—my favorite rock band ever. There was something about it that was slow, confident, almost majestic. My rock had always been more intense. They were relaxed into this groove thing, even if it was up-tempo. Alex lay back, like Ginger Baker always did. Eddie played the way Clapton played, deep in the pocket. He didn't speed up anything. I'd never played with guys like that before.

I called Ed Leffler and told him, "I'm doing it." He told me I was crazy. He thought the Van Halens were nuts and that I was crazy to even think about doing it. Then he went right to work. "Let me see what I can do," he said.

Leffler looked over their situation. Those guys were in bad shape financially—they had made a lot of money, but they had spent it all. They had overhead like crazy. When Leffler found out how much the guys made the year before, he told me I was going to have to take a pay cut to join the band. But once we started playing the music, I knew it was all going to happen.

EDDIE WAS A man of few words. His favorite line was "Yeah, yeah, yeah." All he cared about was getting some rest, having a couple of beers, some cigarettes, and playing music. Eddie wasn't really a driven musician. At one point, making the first album, I grew nervous with his nonchalance, his lack of concern for the whole thing. It wasn't like he was the musical genius telling everybody what to play. Al played the way he wanted, Mike was playing what he wanted. Eddie didn't even know what the lyrics were. He was just concerned about his guitar part. That's all he paid attention to.

When I wrote the song "Love Walks In," his wife, Valerie, was so in love with the first ballad they'd ever done, she made him listen to the lyrics. He got all choked up. "Wow, I never listened to lyrics before," he said. He couldn't sing you one song.

He didn't even know what fucking Dave was singing about. He was listening to his guitar and the groove and making sure that his part was okay.

Mike was my Ed McMahon, always ready to back the play, whatever it was.

Eddie and Al were tight as nails. They didn't get too far from each other. Passed their cigarettes back and forth. One wouldn't light one without lighting one for the other guy. They only needed one match. They never walked into the room with just their own beer. They always had a beer in their pocket for the other guy. It was beautiful unless they'd start fighting, then it was terrible. When they were both drinking, they'd fight at least once a week. I mean, go at it. Fistfights. Mike and I would try to pull them apart. We'd break them up and leave, Al would drive back after we left and they'd go at it again. The next day we'd come to the studio, the windshield would be busted out of the car, the trash can turned over.

I didn't see much of Eddie's wife, but Valerie, wow, what a cool chick. I pulled into the driveway one time and Valerie and Eddie were sitting on the hood of one of Ed's cars, drinking a beer. I thought that was so cool. My wife would never do that. Valerie could hang with the boys. She wasn't around a lot, because she was working. She pretty much always had a gig, some kind of little movie or TV part. She also spent a lot of time at their beach house.

The sense of family ran strong in Van Halen. When I first joined, their father, Jan, was always there, drinking and smoking. Mike Anthony was the most loyal dog on the planet. He was the flag-bearer. From the start, they trusted me and I became the motivator. They loved that and they rallied behind me. It was very family, very close. Us against the world. This is our place. We're working on our record. We didn't argue about nothing. It was a dream come true.

Still, there was a lot of doubt about Van Halen. At the time, no one seemed to have confidence in the band's future. Roth had split with the road crew, the management, and he looked like he was going to launch big-time on his own and leave the band in his dust. People were suggesting that we call it Van Hagar, which was a terrible idea. (Funny though, I didn't know it at the time—Dad always claimed to be Irish—but we were actually Dutch and our family name may have really been Van Hagar once upon a time.)

In spite of the doubt, we already knew it was going to work, because we were the ones in the studio working up the *5150* record and we knew we had some killer tracks. We had "Why Can't This Be Love." However, nobody but us had heard any of this, because we couldn't tell anyone that we were even in the studio together.

Everything had been taking place in secret, because I was signed to Geffen and Van Halen were on Warner Bros. Although we didn't know it, Geffen and Warner Bros. were already butting heads. As Geffen's distributor, Warner Bros. was taking 50 percent of Geffen's earnings and I was Geffen's biggest artist at that time. Elton John hadn't worked on Geffen's label. Neil Young was a disaster—Geffen ended up suing him. Donna Summer didn't have any hits for him. There was me; Don Henley, who had one big album; and John Lennon, who died shortly after giving Geffen his first album, although it did sell millions after he was shot. Geffen wasn't likely to let his biggest act walk across the street just because he wanted to sing with another band.

Leffler and I went to see David Geffen. As we expected, he did not like the idea and wanted to talk me out of it.

"Why would you want to be in that band?" he said. "You're as big as them on your own." He was baffled. He was sitting on his desk, his hand on his head. "I don't understand this," he said. Like a lot of people, he thought David Lee Roth would be an impossible act to follow, and he said so.

After a few minutes of talking it through, he shifted his tone suddenly. "I would never stop an artist from doing what they want to do," he said. "I'm David Geffen. I stand up for the artist. I'm for the artist. Number one, it's about you and your life." He said he would talk to Mo Ostin, the chairman of Warner Bros. Records.

With Warner Bros. over a barrel, Geffen told them he would let them have me for one album, if he could have a Sammy Hagar solo album immediately following. He wanted 100 percent of the solo album and 50 percent of the Van Halen records on Warner Bros. Warner's chief Mo Ostin came to Eddie's 5150 Studios to talk things over. He was, let's say, cautious. He suggested changing the band's name, and he also liked the Van Hagar idea.

Eddie and I powwowed about it and decided, no—we're Van Halen. We loved each other. There was no animosity, no egos, no nothing. They wanted me to be in this band and I wanted to be in it, because we were making the music and we knew we were good. Mo asked if he could hear something, so we put on our instruments, and, while he sat there, we played "Why Can't This Be Love" for him, live and in-person. He put his finger in the air and smiled.

"I smell money," Mo said.

By the time we got the green light from Warner Bros. and Geffen, we were already halfway through the record. After that, we went full-force, and things started happening fast. Eddie and Al had a lot of music left over from what would have been the next Van Halen record before Roth split. They had a lot of sort of semi-formed ideas when I walked in on it. I had to write all the lyrics and melodies. I worked on their jams, picked them apart, and made songs out of them. I was kind of behind the eight ball on the lyrics.

We would jam in the studio for hours. I would have a hand-held mike and headphones and would just sing and experiment.

When something was good, I'd point at the engineer to tell him to make a cassette. I would take the cassettes of the parts that I wanted to keep with me when I drove back to Malibu, which was about an hour's drive. I would drive home, ears bleeding, and listen to the songs.

Both Ed and Al smoked in the studio like chimneys. Those guys would be lighting them up, setting one down, light up another, put it in an ashtray. They would have three or four cigarettes going at one time. They were chain-smokers, lighting a cigarette off the other cigarette, letting the filter burn in the ashtray. Never put them out. Dropping them on the floor. I'd have these terrible headaches when I'd get home at three o'clock in the morning and go straight to the shower, because I stank like cigarette smoke.

One night, on the drive home, I was listening to this tape where Eddie had written the music and noodled the verses on guitar. He was trying to show me the phrasing of the verses, but he couldn't, because he couldn't play the rhythm and the lead at the same time. I didn't get what he was doing. But, on the way home, I heard the rhythm of the thing, and I started singing it in the car. We didn't have a chorus, and I just busted out with it, "Best of Both Worlds." It hit me hard, right when I was pulling in the garage. Bang. The chorus hit.

I went in the shower, but I kept coming out to dry off and write some more lyrics on a notepad. Then I'd get back in the shower . . . and get right back out to scribble down some more. The song came to me like a flood.

I don't know about lyricists. Lyricist Bernie Taupin once told me that it's the easiest thing in the world for him. Once he has a title and a concept, he can just go *bam-bam-bam-bam-bam,* it's done. This song was coming at me like a tidal wave. I couldn't even take a shower. Usually I get pieces that I can remember. I just keep singing them over in my head, and write them down later. I wrote the whole song while I was still taking the shower. I went in the next day and sang it. Everybody was blown away. That was "Best of Both Worlds."

Before I wrote "Love Walks In," Eddie had never really played a real keyboard ballad in his whole life up to that point. With Roth, the closest thing was "Wait" off *1984,* which was a synth track, but it was a rocker. It wasn't a beautiful melody.

I had been reading this book by Ruth Montgomery called *Aliens Among Us.* She claims to be an automatic writer. She just gets a pencil, closes her eyes, and goes into a trance, and the writing comes through her. The book was about walk-ins, aliens who come and take over your body in your sleep. A person can actually not die and still become a whole different person. They wake up one morning and can't remember who the hell they were. I wrote about how love comes walking in and can make you a whole new person. After Valerie forced Eddie to listen to the lyrics, Eddie and I became the closest of collaborators, trusting and loving with each other from that point forward.

BEFORE I JOINED Van Halen, I'd already committed to Farm Aid in September 1985, and we decided that would be the place we would make the announcement. I wanted to really do this big-time. I rented a private plane for my band for our last show. I gave everybody a nice bonus. They'd all bought houses already, but I gave them enough to pay them off, if they wanted. I brought Eddie up to rehearse with us. Eddie and I wrote three songs in two days while he was in town. We rehearsed Led Zeppelin's "Rock and Roll" with Eddie for Farm Aid. It was going to be great. Too bad I screwed everything up.

At Farm Aid, you could only do three or four songs. Willie Nelson, John Cougar Mellencamp, and Neil Young organized the twelve-hour marathon fund-raiser that was broadcast live on radio and television, largely a country show with acts like Kris Kristofferson and Jimmy Buffett. It wasn't a rock show. Dylan was as hard-rocking as it had gotten when my turn came. I was going to open with "One Way to Rock," follow with "I Can't Drive 55," and bring out Eddie, make the announcement that I was joining Van Halen, and play "Rock and Roll." When I went out, right away, I had that stadium rocking. They loved me, were going crazy. I was scoring. It was big for me—ninety thousand people in Champagne, Illinois, one of my biggest regions. I could do two nights in Chicago, two nights in Champagne, two nights in Peoria. Illinois was my state. I was ripping it up when I stepped to the mike to introduce "I Can't Drive 55."

"Here's a song for all you tractor-pulling motherfuckers," I said and instantly they shut down the radio broadcast and turned off the live TV feed. I ruined everything. When I brought out Eddie, we were long off the air and nobody saw or heard a thing. He did a quick little solo, we made the announcement and went into the Led Zeppelin number. That was the first time we played together in public. Eddie was on the plane with us. We all flew back happy. It was a friendly transition from my band to Van Halen.

Many years later, David Lauser went to a cattle-call audition for drummers for Maria McKee, the former vocalist of Lone Justice, this country-rock band signed to Geffen that was going to follow us at Farm Aid. Back then, they'd thought they had the Eagles and Linda Ronstadt all rolled together, but Lone Justice was not to be. A number of years later, the singer was looking for a drummer and Lauser looked like he was going to get the job. He's down to the final interview with the lady herself, and she says, "Tell me about yourself—what have you done in the past?" Lauser says, "I've worked for Sammy Hagar all my life." She gets up and walks out of the room. I guess there was still some animosity about that. I didn't mean to do anything wrong.

When I first went down to join Van Halen, I moved into a rented house, but then a place in foreclosure came up for sale next door to Eddie and Valerie in Malibu. They lived on a bluff, a little house in the middle, and my house. It was all brand-new, but the contractor went under and the bank had it. I made them a ridiculously low offer and got the house. I moved in next door to Eddie. It was amazing that a place would come up on Broad Beach Road, one of the most desirable spots in North Malibu, especially at bargain-basement rates. It seemed like karma.

Betsy was grooving. She dug the house and she liked the beach. She didn't care about Malibu, but there were horses right down the street. You can rent a villa in the South of France for what we paid for her horses at the Malibu stables, but she was happy. When I joined Van Halen, it shook up Betsy bad, at first. She'd been ready for me to slow down and get out of the business. She wasn't ready for me to start over with a totally new band. But when the house came up in Malibu, she started to see things differently. I would go to work every day and come home at night. We were recording the album and rehearsing for the tour. She was living in this beautiful beach house. She had her roses in the garden and her horses down the lane. She drove either her brand-new Jaguar or the Land Rover.

I used to drive to the studio with Eddie every day. He and I had the cars. We'd take either the Ferrari, the Lamborghini, my E-Jag, or my Cobra. On the way into the studio one day, we drove past a dealer and saw an E-Jag sitting there. I stopped. I went in and got my business manager on the phone, handed the salesman the phone, and the two of them put the deal together. I went out to get in the car and drive to the studio. I slid in behind the wheel and, wait a minute, what's with the seat? I'm not that big myself, but whoever drove this car before was one short dude. Whose car is this? It belonged to Ronnie James Dio. I loved that.

Eddie and I did crazy shit like that. We'd race home, me in one of my Ferraris, Eddie in one of his Lamborghinis, driving 140, 150 miles an hour. He would always be drunk.

Every day, we'd go from two in the afternoon until past midnight, unless Al passed out. Al was a bad drunk, but Eddie used to nurse his beers. He'd always be drinking, but didn't get all fucked up. Al would get fucked up, puke, pass out. You'd have to slap him around, let him rest for a couple of hours, get him up, and bring him back. We would limit the amount of beer he could have and he would duck out for a pack of cigarettes, run downtown, buy a bottle of vodka, and drink it in the store.

Their father liked to drink, too. If we were in the studio at two in the afternoon, it wasn't like Eddie got up at eight in the morning, more like noon. I'd get to the studio and the three of them would be sitting there drinking, having gone through a couple of six-packs already. Their dad, Jan Van Halen, was a great guy. I felt close to him. He was a sax player. He liked my chops, liked that I could sing.

But those guys drank. Al was a drunk like my father. He couldn't stop. He drank until he passed out, woke up, and started over again. He would find people in bars and offer them money to put out a cigarette on their arm or shave their head, while he videotaped the whole deal. Completely nuts.

When I first made their scene, they were still laughing about Al's birthday performance. Legend has it they'd all gone to a Benihana. He was already drunk when he got there. It's his birthday. He's drinking hot sake and everything else. He gets up on the table, takes his shirt off, and starts to dance—right on the hot grill. The guy just finished cooking dinner on it. Al pulls his pants

down to fuck with the people in the place. He trips himself because he's got his pants around his ankles, and lands on the grill
on his back. *Ssssss.* He can't get up. He flops over like prawns.
Sssss. Ahhh, ahhh. *Sssss,* ahhh! He couldn't do anything. He was
on the grill. They had to pull him off and, of course, take him to
the hospital. He had burns all over.

Like Leffler said, these guys were crazy, very high-maintenance,
but good-hearted. Another time, during the *5150* sessions, we were
waiting for Claudio to bring back one of my cars around two in the
afternoon.

"I bet you I could shotgun ten beers," Al said.

He's got ten talls of malt liquor. "I bet you a thousand bucks,"
he said. Al's a betting man. One time he lost his BMW to me on a
bet. I made him pay up, too, and gave it to our tour manager for
Christmas. Al was a great guy, but just a total fuckup. I was not
betting him.

"Oh yeah?" he said. "Watch this."

Michael Anthony was standing there with me. *Pow, pow, pow,
pow*—Al opened them all first and then drained them, one at a
time. How can you even hold that much in your belly? I thought,
"Oh, no, this day is over." Al walked out into the driveway, big
belch, and grabbed a broomstick. "My dad used to do this," he
said. "You ever see anybody do this?"

He was standing on the asphalt in the driveway, holding the
broomstick out in front of him with both hands. I looked at him
and wondered, "What the fuck's he doing?"

He's going to jump over the broomstick while holding it in his
hands.

At that moment, Claudio pulled around the corner in my car.
Just as Claudio could see everything clearly, Al jumped over the
fucking thing, caught his feet, of course—drunk, just shotgunned
ten cans of malt liquor—and went down, face-first. He didn't let

go of the broomstick. We had to take the broom out of his hands. He hit the ground facedown and lay there, out cold. Claudio jumped out of the car, screaming. "Call an ambulance," he said. "Oh, my God, he's dead."

He did hit hard. The momentum of trying to do it catapulted his head into the asphalt. When we picked him up, he had a pizza face. They took him away in an ambulance.

I went home, took the next day off, and, the day after that, I came in. Al was lying on the couch, his head wrapped up like a mummy. I laughed at him so hard, but he couldn't laugh—that only made it hurt worse. He really did some pretty good damage to his face. That was just starting to make the record. You could only imagine what the tour was going to be like.

When it came time to actually record the album, we needed a new producer, because Ted Templeman, who was producing David Lee Roth's solo albums, had supposedly been bad-rapping us to Warner Bros. behind our back, so we weren't going to work with him. Despite Mo Ostin's positive reaction when he'd heard us at 5150, and the fact that they weren't paying much for it, Warner Bros. was hardly enthusiastic about the project. I suggested Mick Jones from Foreigner. I'd known him from the Montrose days, when he was still with Spooky Tooth. So Mick came on board and produced the album with us.

One day toward the end of the project, Mick and I were walking on the beach when he turned to me and said, "Give me one more song." That was "Dreams." He just sort of pulled it out of me. I didn't know what key the song was in. I started singing in that register. Mick got really excited and helped me learn how to sing in this supersonic range that I'd never done before. He pushed me into a range that was an octave above where I normally sang. Mick got me to do things I didn't know I could do.

We cut the album quickly, no more than a month, but we got

hung up mixing. It took longer to mix than we expected, because Eddie's studio really wasn't a great place to mix. We would do the mixes, take them home, and not like what we heard, so we'd have to do it all over again. We had to cancel dates we planned at the start of the tour in Alaska and Hawaii. We wanted to start in some remote place, because we were really concerned about how the people were going to respond to the new material.

There was also the issue of the Van Halen catalog. I told the guys that I didn't want to be in a cover band. I was not going to go do any shows until we had an album, and when we did, I didn't want to play too much of the old shit. They were totally down with that. We all decided to go out and make a stand.

In the end, we were so late delivering the tapes that the album couldn't come out until a week after the first scheduled date on the tour. Rather than start in some faraway place, we began it all in Shreveport, Louisiana. Even though the album was late, we went ahead with the show, because it had already sold out and we didn't want to cancel or postpone it. The single "Why Can't This Be Love" was out on the radio, so people had heard something, but they hadn't heard the album.

We went to Shreveport, Louisiana, to do that first show, March 27, 1987, and I was a wreck. I don't think I've ever been more nervous before a show. We came out and opened with "One Way to Rock," one of my songs. The barricade went down. The audience went crazy. It happened in an instant, a flash. It was killer. We knew we were on top of the world at that moment. ·

The ironic thing about that date is that it had been predicted a couple of years before by a psychic named Marshall Lever. I met him through this acupuncturist I'd been seeing and went to visit him at his home in Sausalito for an appointment sometime during the recording of *VOA*. This was after I had severed my ties with the girlfriend and was happy with the new baby, Andrew, but I still knew things weren't right between Betsy and me. I felt the need to talk to someone. I needed some spiritual advice.

His wife met me, this red-lipped woman, very goofy, who showed me into the room. This heavyset gentleman walked in, sat down in a rocking chair, leaned back, and closed his eyes. His wife asked if I wanted to record the session and slipped a cassette in a tape recorder. His dog followed them into the room, lay down on the floor, and started snoring. Over the years, I've been to see this guy twenty times and this was the routine. That dog is snoring on every one of my tapes.

He started by telling me that I was involved in a relationship that I was just finishing. "She was your sister in your past life in Greece," he said. "You were separated when she was nine and you were eleven, and your parents were killed in a boating accident in the Greek islands. They put her in a convent and you went out on a fishing boat and never returned. You never saw her again, and you missed her. When you saw her and you smelled her"—he's talking about the smell, this chick drove me crazy with her smell—"when you smelled her, you realized who she was and you didn't ever want to be away from her again."

He went on to tell me about Betsy. "Betsy was also your sister in a past life," he said, "and you lived in Spain. Neither one of you ever married. You were in love but you never had sex because she was your sister and you lived together your whole life. Betsy was your big sister. Your mother died giving birth to you. Betsy cooked for you, just like your wife, but you never had sex even though you were madly in love. You were an instrument-maker named Crulli, C-R-U-L-L-I." He spelled it out. "And your instruments can be seen in a museum in Barcelona."

He turned his attention to Betsy. "When you met your present wife," he said, "you had an extremely strong sexual relationship and it's really what keeps you tied." That's really what we really had. Our sexual relationship was fantastic, even twenty years into it. How did he know all this?

"In eighteen months," he said, "you're going to go on a brand-new adventure, very much like what you're doing now, but different. More powerful, bigger, more like 'this is it.'"

That's when he gave me that date. He said it was going to start on that date.

I refused to sing "Jump." It was just hard for me. I write my own songs. "Jump" was tough for me lyrically—"Can't you see me standing here, I've got my back against the record machine, you know what I mean, you know what I mean? I might as well jump." That was hard for me. I couldn't sing the song with any heart and soul. I've got to sing something that I mean.

"Hey, hey, hey you, who was it? Hey, baby, how you been?" I just couldn't sing that shit, great as it was. The first night, in a moment of panic, I pulled a guy out of the crowd to sing it. The audience went nuts. The band thought it was great. When the guy got to "I might as well," I'd spring in the air like a maniac. It worked. We kept it. On the entire tour, I sang "Jump" maybe twice.

Before I joined the band, Van Halen didn't have a particularly tight show. Roth would talk. They'd do another song. Ed would play a twenty-minute guitar solo. They would do another song. Roth would talk some more, another song, Al would do a drum solo for thirty minutes. On the 1984 Tour, they told me they were doing eight songs in a two-hour show. And they ended every song the same way. They had the classic heavy-metal ending—four crashes, a crescendo . . . one, two, three. At the end of that, Al would usually do something, smack this, clang that, just because he was quirky. I decided we needed a new ending.

"Great idea," said Eddie, as always. So, pounding beers, Al learned a new ending. Good. The next day at rehearsal, back to the same old ending again. If he learned it, he learned it for one day maximum. Nothing stuck. We kept the same ending on the tour.

On the road, the crew worked around Al carefully, trying to figure out ways that he wouldn't pass out during the show. Al would sleep right up to the time before we went onstage. I would come to the dressing room from the hotel—I never did the sound checks to save my voice—and the two of them would be asleep on the couch or in a chair. They never went back to the hotel for their naps. Everybody tiptoed around them.

"Shhh, let them sleep," they would say. "Don't wake them up or Al will start drinking too soon."

They would wake up Al about twenty minutes before show-time. There was always a case of tall Schlitz cans. He would shot-gun three or four beers and get his buzz on. He would walk out onstage with a couple more cans in his hands, pound those, and drink the rest of the case during the two-hour show. The crew would put out these big rubber trash cans for him to piss in during the show. After practically every song, he'd piss in the trash can, pound a couple of beers, and start playing again. Sometimes he'd really be fucked up. In the middle of a song, he'd just get up off the drums to take a piss or chug a beer. Eventually he started wearing one of those helmets with beer holders on the side, and straws. At the end of the tour, he needed some help.

This was the golden era of arena rock. I had been doing arenas since 1982 and *Standing Hampton*. I was raised on arena rock. Montrose opened for everyone in arenas. I never played nightclubs and theaters. I wouldn't even know what it was like. I was used to going out with the big moves, hands as far as you could stretch them, running across the stage, jumping as high as you could to get to those people at the back of the giant arenas.

Van Halen was the classic arena-rock act. At the end of our run, arenas had gone away. People started playing amphitheaters for more money. Arenas were smaller and more expensive. You couldn't bring a giant production into the amphitheaters. When we first started doing arenas, Ed Leffler and I came up with a way to streamline production and stage design so that we could sell an extra two thousand seats in the back, behind the stage. Those seats were pure profit. We didn't put a canopy on top of the lighting, so people could see the stage. We raised the PA system. We learned all the tricks and invented a few of our own. When we were running through arenas after *VOA,* we made more money than the other bands, because they weren't selling those last twenty-five hundred seats. When I joined Van Halen, they had been draping off the back of the hall, cutting their capacity in half and walking away with a few thousand dollars. On the 5150 Tour, we designed the stage so we could be seen from everywhere.

The arenas were so big and grand, and had roofs all the way to the back. You could extend your production as far back as you wanted and you could have as many as fourteen spotlights. When you came out, it was big-time rock. It was loud. It was inside a building and sound didn't just disappear like it does outdoors. The sound was contained in the hall. It was massive and thunderous and the audience felt it in their chests. You could darken the entire building and, then, *pow,* hit these four little guys up there with four massive spotlights apiece. Arena rock is how rock stars became rock gods.

For the 5150 Tour, we built this giant stage with steel gratings that went up to another stage about eight feet higher, which went all the way to the back. That way, I could work the crowd in the back. We had an eight-foot lift, where the drum riser was, that was like another stage. I would go up there and Eddie would go up there. Mike would go up there. You could be closer to the rear sections than the front row, even exchange high fives with the crowd.

We had two other platforms on each side. Our stage was

massive. We had these trusses of lights that I had been using since the Three Lock Box Tour, with catwalks on them. It came down as an X across the front of the stage at a point in the show, and I went up there and out over the audience twenty or thirty feet above their heads. We carried the show in fifteen trucks, a huge amount of production, some special effects, but mostly sound, staging, and lighting.

I was one of the first to use the headset mike so I could run around all over the place. We were all wireless. We used to come out on this massive stage, and we wouldn't see one another again for ten minutes. Eddie would be running one way. Mikey would be jumping off in another direction and I'd be somewhere else. Only Al was stuck where he was. Sometimes I'd put my hand up over my eyes so I could see where Eddie was on the stage. We kept our monitors out of sight, under grating on the stage, so the stage was clean except for the tall amps. And they were loud. If you went in front of either of the amps, you'd better hold your ear. Van Halen played loud. The PA had to be so loud because it was coming off the stage that loud. That was arena rock—Led Zeppelin, Black Sabbath, Deep Purple, Rush, Van Halen. Rock stars.

On the tour, there was a former Playboy bunny from California hanging around, who used to see one of the other guys in my old band. Somehow she hooked up with Leffler, although she had always been after me. She was good-looking, but there was just something about this chick that was not to be trusted. She saw my name on Leffler's rooming list and came knocking at my door in the middle of the night in Detroit. I answered the door without any clothes—I sleep naked—and she pushes the door open, throws me on the bed, and starts blowing me. That's kind of tough to get up and walk away from. "Son of a bitch," I was thinking, "I'm fucked now." And sure enough, I was.

About ten days later, Leffler gets the phone call. She's pregnant.

I smelled a setup. I was so pissed off. Betsy would commit suicide. We hired an attorney and started dealing with her. I knew it was not my baby. It was extortion.

She wanted an apartment in New York and anything for that kid that my children would have. I didn't want to pay a penny, but Leffler convinced me the smart thing to do was give her the money until the baby was born and see what happened at that point. She was living with her boyfriend, a musician in New York, in the apartment when she had the baby. She called Leffler from the hospital. "Tell Sammy to call me," she said. I didn't want to talk to her, but Leffler talked me into it. She tells me the baby is so cute, looks just like me, she's madly in love with me, she's so sorry, shit like that.

A couple days later, Leffler gets another call. The baby died. I don't believe that she ever had a baby. She may have had an abortion early on. Marshall Lever, my psychic with the sleeping dog, told me about it. "It's not your baby," he said. "She's living with her boyfriend in New York. She has a boyfriend that's a musician and this is probably an extortion case. Don't worry, just relax, and once she has the baby, it's all going to go away."

I never heard from her again. Obviously, it wasn't my baby, and they knew it. They just extorted me as long as they could. No one ever saw her again.

Three weeks into the tour, we were sitting in Atlanta and Ed Leffler called a meeting.

"*Billboard,* number one," he said.

It was the first number-one record for any of us. The album sold 600,000 the first week and another 400,000 the next week. It was on fire. It went platinum faster than any record in Warner Bros. history. Every one of our records did. When I was in the band, Van Halen was a huge, quick seller. Every album went to number one. It was an un-fucking-believable run.

✴8✴

MONSTERS OF ROCK

As soon as we finished the first tour, I had to make the solo record for Geffen. That was the deal he'd made with Warner Bros. We were worried about Eddie and his drinking and drug problem, but first we had to deal with his brother. We put Al in rehab as soon as the tour was over. His wife staged an intervention. I didn't even know what an intervention was.

It was hairy. I cried. It broke me down and I wasn't even the guy under the gun. They went and got him out of bed, six o'clock in the morning, before he had another drink. He was getting up at four o'clock in the morning, chugging a bottle of vodka, and going back to bed. He wasn't a sipper. He wasn't a nurser. He just plowed himself to the point of passing out.

We put him in a hospital. He took the oath and never drank again. I love Al. He is the strongest guy, but weird. He's a chain-smoker, but he'd quit smoking every Monday, for the one day, just to torture himself. Al's the kind of guy that I'd call every day, just to bullshit.

Once Al cleaned up, Eddie didn't have anybody to drink with. Al still smoked. Al would drink coffee and Eddie would drink beer and do a few other things. It was Leffler's idea to have Eddie coproduce my solo record and play bass. He always played bass with Van Halen, two or three songs on practically every record. He was a great bass player. Eddie's a great musician, period.

So off we went into the studio, with Jesse Harms on keyboards, David Lauser on drums, and Eddie on bass. I played guitar. That way, we'd keep Eddie busy. We cut that record at these brand-new studios A&M Records built in Hollywood, where Tom Petty, John Cougar Mellencamp, and Stevie Nicks all had been working. Pink Floyd was in the room next door, without Roger Waters, doing "Learning to Fly." Eddie and I would ride in together from the beach every day in a different car. I had about seven Ferraris down there. The Pink Floyd drummer, Nick Mason, is a big Ferrari collector. The guitarist David Gilmore owns Ferraris, too, but he's not in Nick Mason's class as a collector. Mason owns one of the original Ferrari GTOs, a car probably worth 30 million bucks. They didn't have their cars with them, so they'd be waiting on the sidewalk every day to see what I was going to be driving. David Gilmore is one of my all-time guitar heroes and it was really cool, having those guys admiring my cars every day. I was showing off. After I'd run through all the cars I had down south, I sent Bucky back up to Mill Valley, to swap out a couple more cars.

Pink Floyd was auditioning drummers for a shuffle they couldn't nail, even with their drummer Nick Mason there. They had Omar Hakim trying out, fresh from Sting's band, but they didn't use him on the track. I did. He overdubbed drums on a couple cuts on my album. Pink Floyd was so particular about that shuffle, they were still working on it by the time I finished my entire album.

MTV did a whole "Name the Album" promotion, because I couldn't go on tour. I was just going to call it *Sammy Hagar,* but some fan submitted the title *I Never Said Goodbye,* with a note saying, "Sammy's left his solo career but he never said goodbye."

The record went platinum immediately. "Give to Live" and "Eagles Fly" off the album were big hits. I did a three-week promotional tour around the world—San Francisco to Japan to Germany, came home and went straight into the studio to start the new Van Halen album, *OU812*.

As soon as I came back home, I flew down to Los Angeles from San Francisco. Eddie and Al met me at the airport. I hadn't seen them for a couple of weeks. I was really happy to see them. When we got in the car, Eddie and Al lit up cigarettes in the front seat and snapped a cassette in the player.

"We want you to hear something," Eddie said. They played me the keyboard part for "When It's Love." I was covered in goose bumps. That was almost the inspiration for the whole album. We knocked that song out and knew we had something.

The songs were not my best stuff lyrically. "Black and Blue" was kind of quirky, cool groove and phrasing, but the lyrics were a little too eighties. "Source of Infection," ugh. The last song we wrote was "Finish What Ya Started." It was toward the end of the project and we needed another song or two. Ed was the best at taking an idea you gave him and turning it into something special, something unique. I told him we should do something like the Who's "Magic Bus," something with a lot of rhythm and acoustic guitars. Van Halen hadn't really done anything with acoustic guitars.

I was in Malibu, lying in bed with my wife, about to get some, when I heard Eddie outside my door. Not even my front door, but the beach entrance directly under my bedroom deck. I could see him out there, cigarette glowing in the dark, no shirt, acoustic guitar around his neck, bottle of Jack Daniel's in his hand.

"Ed, what?" I said.

"Come on, man. I've got this idea," he said.

"Ed, it's two o'clock in the morning," I said. "I'm tired."

"The old lady kicked me out," he said. "Come on, man. Let me in."

I went down. Betsy was pissed, but what could I do? He was my best friend and creative partner. She turned out the light. I wouldn't let him in, because he's got his cigarette, so we sat on the porch. He started playing me the riff for "Finish What Ya Started" and right away, I got excited. I went and got my acoustic and started doing my Tony Joe White thing. I was still thinking about going back upstairs and getting laid and started singing, "Come on, baby, finish what you started." It never happened. Eddie and I saw the sun come up and I threw him out, but we had written just about the whole thing. I was imagining what was going to happen to me when I went back upstairs. That song is about unfulfilled sex.

It took a while to do *OU812,* more than it probably should have. Sober, Al played different. It was really weird. He wasn't as good. When he was drinking, Al was always a radical drummer. He'd hit hard and do crazy fills. Drunk, he was off the hook, but sober, he was a lot more conservative. His timing was better, but he wasn't as radical. In the end, it was easy to put aside things like that, because everybody was still getting along really good. The only problem was that I was starting to burn out, getting a little crispy around the edges.

Betsy, on the other hand, was losing it. I'd come off the VOA Tour, jumped into *5150,* did the album and the tour, did my solo record, went around the world on the promotional tour, and straight back into the studio for *OU812.* And now we were going back on the road. She was crying, depressed all the time. Life was not good at home. I was not ignoring it any longer. Now I was concerned.

I went to Ed Leffler and suggested we do stadiums. We were in the biggest band in the world, and as it was, we were selling out four nights in arenas anywhere. Stadiums would allow us to go out for a couple of months instead of a year, and if things went right, we'd end up making way more money. Leffler thought it was a gamble, but worth taking.

We put together a bill. Kingdom Come was a new band from Germany that everyone thought was going to be big. The Germans sounded exactly like Led Zeppelin, but they told interviewers they never heard of the band. So they were history before they were out of the blocks. But they opened. Metallica, who went on second, had been my choice. They were a new band from the Bay Area, young and hip. Dokken, the third act, were ready to break up when the tour started, and following Metallica every night finished them. They broke up at the end of the tour. The Scorpions were also on that tour and they meant a lot. We called the tour Monsters of Rock and did twenty-seven shows between Memorial Day and the end of July 1988.

We had fifty-six trucks, three complete stages, and production systems that had to be put up all at the same time so that we could go out and play three shows a week. It cost $350,000 per show and it was easy to lose money. The shows broke even at forty-four thousand tickets and the tickets were expensive.

But the fact that *OU812* went straight to number one made selling tickets easier. It went on to sell 4 million records and had the big hits "When It's Love" and "Finish What Ya Started." Meanwhile that tour made a lot of money, even if it didn't do so hot in some places. We sold out two Detroits. We sold out two Texxas Jams. Sold out Candlestick Park in San Francisco. We sold out a couple of New Yorks. But, in Miami, we did around 25,000 people, and a hurricane blew in and ripped the concert to shreds. We had to stop in the middle. Big loss. We played a couple of other markets where it didn't do so good and we ended up giving a lot of money back to the promoters. It wasn't a disaster, far from it, but it wasn't the home run we expected.

On the first song, opening night, I fell on a metal step. Everything had been running late and we didn't have a full production run-through the day before like we normally would. The stage was done barely in time for us to play on it. I didn't know the stage. I tripped and landed right on my tailbone, spent the whole night after the first show in the hospital. I was getting steroids shot into my ass every day for my broken tailbone. I was getting massages, and I had to sit on ice packs. I had a grapefruit on my lower vertebrae.

On days off, I flew home to see my doctor. On one of those trips back, I hobbled into the waiting room and there was Miles Davis sitting in a chair. The doctor opened the door and out comes Sting and his wife, Trudy. "Look. It's Sammy Hagar," Sting said. They left and the doctor said, "Sammy, have you met Miles?"

Fucking Miles Davis sitting there, dressed like a woman in his shiny, crazy-ass clothes, skinny corn rows in his hair. Miles puts his hand out, without standing up, and as I reach to shake his hand, he reaches out with his other hand around my forearm, and he pulls himself to his feet. He's using me to get up out of the chair. I almost went down with him.

I had an ear infection. I had a sinus infection. I had a broken tailbone. That whole tour, I couldn't sing because of my sinuses. Every night, I couldn't sing. It was too big of a deal to cancel because it was so expensive.

We got to Texas, standing out in front of sixty thousand people at the Cotton Bowl, and I couldn't sing. I was about to cry onstage. Texas was my country. I owned Texas. I was a headliner there long before Van Halen. I stopped in the middle of the first song.

"I can't sing," I said. "I promise you, Van Halen will come back and do a free concert for Dallas."

We cut the show short, and the brothers went nuclear on me afterward. They crucified me for that. It was three years before we made good.

Earlier that day at the Cotton Bowl, we'd had an opening band called Krokus, who were handled by Butch Stone, a Southerner I knew from Montrose days when he managed Black Oak Arkansas. They planned to broadcast their set live on the radio, but Ed Leffler pulled the plug on them. Butch was really upset. There was a screaming match. That night at the hotel, after the bar closed, somebody jumped Leffler in the elevator and beat the shit out of him, knocked out a couple teeth, broke some ribs, really busted Ed up. He spent two weeks in the hospital.

Still we rolled along. I told Betsy I was only going out for one month, but we came back and Leffler started talking about going back out to hit the other markets. The band had me in the backseat of a car coming back from some press function—Ed and I were drinking and doing coke—and they started drilling on me. With Al sober, he couldn't sit around the house all day.

"I need a life," he told me.

Eddie liked to be on the road, too. He and his brother would have played seven nights a week, if they could. They were really pressuring me.

"I just can't do it," I told them. "I love you guys, but I'm going to lose my marriage. I'm going to lose my family. It isn't worth it to me." It wasn't like a fight, but we were bickering.

"We got you in the band, we thought you were committed," Alex said. "If we would have known this, we would have gotten somebody else."

Alex sober was a head-tripper. Drunk he was just tripped out. Eddie would always back him, no matter what. You could not come between those two guys in a million years. They're not just brothers. They came from another country, didn't speak English when they got here, and were tight. They'd fight like a cat and dog, but don't get between them.

The brothers decided that Mike Anthony wasn't making enough of a contribution to continue earning a full share of the music publishing. Truthfully, Mikey didn't write. Ever. He basically played on bass what Eddie would tell him. He was a quick study and he would add to what he was given. Mike was a creative bass player. He had an incredible background voice that made a big difference in the sound of Van Halen. But the brothers needed money.

Al was almost broke. We were making tons of money, but Al was many million dollars in debt. Leffler helped consolidate his bills and arranged a five-year window with him just paying interest. He had a little real estate crash. He'd bought a $2 million house that he sunk even more millions into for rock-star junk like a rubber room, and then sold it for a big loss. He was making really bad deals.

We held a meeting and took a vote on reducing Mike's partnership to 10 percent. Leffler and I voted no, but Mikey sided with the Van Halens and voted against himself, 3-2.

"I understand what you're saying," he said, "and I'm okay with it."

I knew going back out was inevitable. We spent fall 1988 on the road. We didn't do as hard a tour. We went back to some of the markets that we'd canceled. We went to some of the secondary markets we missed, trying to keep the record alive and make up for the bad press from the Monsters Tour. Everyone thought it bombed. It really didn't. We made a lot of money, and we took care of everybody. No one lost money. But attendance wasn't what it was supposed to be.

We were also doing whatever we could to pump album sales, which were, again, great, but not what they were supposed to be. As a result, we started making music videos. For *5150,* Van

Halen made no videos. Ed Leffler figured he didn't want to compete with old Van Halen videos. Also, Warner Bros. refused to pay for the videos and they could be expensive. When we put out the first single, "When It's Love," from *OU812* without a video, too, Warner started to freak out. They thought if we had put out a video with *5150,* instead of selling 7 million, we would have sold 10 million. We were already out on tour when the second *OU812* single, "When It's Love," came out, and Warner finally agreed to pay for a video. Leffler insisted it be a performance video. No acting. It needed to be shot on one of our off days from the tour.

They sent the Warner jet for us, flew us to Hollywood. We spent twelve hours shooting the video and I blew out my voice. At two in the morning, they loaded us back on the jet and sent us back to the next town on the tour. They had shot some B-roll with an actor and actress, and she was around for the shoot, wiping down the bar in the background of the shot or some such. We really connected, but I never even got the chance all day to take her in the back room.

The second video was "Finish What Ya Started," where director Andy Morahan wanted to shoot everybody individually. He wanted to make sure everybody looked good all the time and you can never do that with four guys all at once. Someone always gets caught with his eyes crossed, his finger in his nose, his chin doubled. Leffler insisted on another performance video again. They shot us in high-contrast black-and-white against a white backdrop. We each played the song for about three hours. When we walked in, the place looked like the men's department at Macy's, there were so many racks of clothes. We wore what we were wearing. We didn't let the stylists touch our hair or do our makeup. A lady put a little powder on and that was all. We would see all these dolled-up, blow-dried bands on MTV and laugh at the assholes. We weren't going to do that.

✳ ✳ ✳

WE WENT AND did those extra four months' worth of shows, but by the time I came back home, my marriage was coming apart. Betsy was getting a distant look in her eye. I started worrying about her. Al's marriage was also falling to pieces, and the four months on the road didn't do that any good, either. With Al sober and Eddie still fucked up all the time, they were fighting like crazy. We broke up fights constantly. Over stupid things.

"Hey, Ed, you got a cigarette?"

"Hey, Al, why don't you buy your own fucking cigarettes?"

"Hey, fuck you, man."

"Well, fuck you."

Boom. They'd just go at it because he asked for a cigarette. There was always tension between them. When they didn't want us to know what they were arguing about, they would shout at each other in Dutch.

We didn't sell out everywhere on the second leg. We still did big business, but we were doing eight thousand out of twelve thousand seats, and twelve thousand out of fifteen thousand, instead of two or three sold-out nights. Some cities, it was always there, others we had to work harder for.

Even though the *OU812* album sold 4 million, people thought it wasn't successful, since *5150* had sold 7 million, almost twice as many. Maybe the honeymoon's over. Big pressure on me now. Now we've got to do a great record and go out and tour the rest of the world. We hadn't toured the world yet. I was resisting. I didn't want to go. That's when Betsy had her nervous breakdown.

It all went down shortly after I had bought this airplane, a Merlin 3, and started spending a lot of time in Mexico. We had used it a few times, and I decided to buy it, a seven-passenger twin turbo prop, small and fast, the biggest plane you can have with one pilot. I was stretching a little bit, buying this plane. I could afford it, but it was a very big luxury. I redid the interior with all this white and beige, cream-colored leather and suede. I put in a flushing toilet. I had a seat that folded down into a double bed so that the kids could sleep.

The first trip we took on the new plane to Mexico, Betsy kind of freaked out. We arrived and she was all nervous. She couldn't sleep and started shaking, getting really bad anxiety and panic attacks. With us was Bucky's ex-wife, Joelle—they were divorced by then. Bucky had been living with their son, Benny, and Joelle was working for us as a nanny. We had to leave Mexico, because Betsy was feeling terrible. We got on the airplane. We were cruising at twenty-six thousand feet, halfway up the middle of Baja, when she started screaming and pulling her hair out. She tried to open the door and jump out of the plane.

Aaron wasn't with us. Andrew was. He was a baby, about two years old, sleeping. Joelle had me hold Betsy down, and I yelled at the pilot to land. He told me I did not want to land in the middle of nowhere in Mexico with a crazy woman on the plane. "You've got to let me try to get to San Diego," he said.

I had some booze on the plane, and Betsy was willing to drink it at this stage. She never drank, did drugs, or anything, but she was freaking out so badly she couldn't breathe. I poured tequila down her throat, while Joelle held the baby's ears.

Betsy finally calmed down. She passed out. She was drunk. When we got to San Diego, I hired a car to drive her to Malibu. We were originally on our way home to San Francisco, but we had our house in Malibu. I took Andrew with me on the plane and she and Joelle went by limo to Malibu. Joelle was her good friend. She really took care of her. By the time she got home to Malibu, Betsy was in bad shape.

I called the doctor and he came straight over. He gave her shots that calmed her down. He told me she was having a panic attack. I had never even heard of a panic attack. There was a psychiatrist who lived across the street. He started coming over to our house for an hour or two every day. He put her on Prozac, Xanax, every kind of antidepressant, mood-enhancing drug you can take. Betsy had never taken drugs in her life.

I had twenty-four-hour nurses. I couldn't leave her. I couldn't even go to the grocery store. If I told her I was going to the store, she would fall on the floor and curl up in a little ball. I was the only one who could bring her out of these trances. We'd try to pry her open, get her to get off the floor, put her on the couch. Her pupils would dilate and she stared straight ahead at nothing. Nobody else could talk her down. I had to be around.

I held a meeting with the guys in Van Halen. I told them I cannot go in the studio. I cannot go on tour. I've got to get my wife healed.

WE TOOK A year off in between the OU812 Tour and *For Unlawful Carnal Knowledge*. Ed Leffler and I both believed that time off was good for a band who had been in everybody's face so hard for so long. Promoters, record companies, everybody involved with making money off you will tell you to stay out there while you're on top. They tell you the public forgets real fast.

They don't. The Stones prove it every time they go out. Pink Floyd waited as long as they wanted to go back out. It's the bands that tour, tour, tour that go down. Pretty soon nobody wants to see them anymore after they've seen them twenty times the last two years.

Eddie didn't like being home, not doing anything. I was watching over Betsy every night, every day. Ed and Al were really putting pressure on me. They would be writing and come up with something cool, like "Poundcake." I would hear what they were doing and really wanted to be working on it. I'd go in for a day or two.

We finally went in the studio to make the next album, *For Unlawful Carnal Knowledge*. We wanted that Led Zeppelin sound, so we went to Andy Johns, the British engineer who worked on the original Led Zeppelin albums. Too bad he was so trashed. He and Eddie were fucked up most of the time. We would start around noon, because I wanted to be home by dinnertime to help Betsy.

Making that album was kind of like pulling teeth. Betsy was under twenty-four-hour care and the situation at home was tense. I was trying to stretch out the process, not because I was being lazy. I wanted the album to be great and I wanted to be there as much as possible, but with all the problems at home, there were days I couldn't go to the studio. Eddie and Al were there night and day, every day. They started hammering me a little bit about not being there. Eddie would call up and say, "We need you—when are you going to be able to come in?" I'd go in for a couple of days, sometimes a week straight, and sometimes I couldn't come at all. I'd bring music home and listen. But it's really hard to concentrate when your wife is curled up in a ball on the floor, crying.

Andy Johns was a disaster, but Eddie protected him. With Al sober, Eddie needed a new partner in crime and that was Andy. He was bombed a lot of the time. He crashed his car into the studio wall. I was not happy with the situation. There started to be a little tension. Then Andy erased one of my vocals. That was it. I wasn't working with the guy anymore. I stormed out of the studio.

"Samster, come on, mate," Andy said. "Just one more time."

"Fuck you, Andy, that's it," I told him. "I'm done with your ass."

It was hard enough getting a vocal with him because he was so disruptive and didn't pay attention. I threw Andy Johns out and I brought back Ted Templeman, who not only produced Montrose and my *VOA* album, but handled the early albums with Van Halen when Roth was still in the band. Eddie and Ted didn't get along, and Ted had bad-mouthed us when we'd originally formed before *5150*. Still Ted came in and did all the vocals with me on *For Unlawful Carnal Knowledge* and helped with the mix. He saved the day for me.

The title was my idea. Around this time, the Florida rap group 2 Live Crew raised the issue of censorship in the record business. People kept asking us to take a stand. I thought, not to get all political, that we should call the next record *Fuck Censorship*. Leffler thought chain stores would refuse to carry a record with that title. Van Halen, the biggest band in the world? Every album number one? They're going to do something, stuff the thing in a brown paper bag, something, anything. They're going to figure out some way to sell the record, those greedy bastards.

At the time, Ray "Boom Boom" Mancini, the former light-weight champ, had been training me. He came over a couple of times a week and put a hundred-pound bag of sand on my back and made me run up and down the seventy-seven steps that went down to the beach from my house. I was playing him some of the stuff, and he asked what we were going to call the album. "*Fuck Censorship*," I said.

"Oh, wow, man," he said. "For Unlawful Carnal Knowledge."

I'd never heard that before, so he told me that in medieval days, when a woman was caught cheating, they'd lock her up in the town square, and hang a sign around her neck that said "F.U.C.K."— "For Unlawful Carnal Knowledge."

"My mom told me that when I was in school," he said. "First time I said 'fuck.'"

"Poundcake," the first single from the album, was a great video, number one on MTV, Rock Video of the Year in 1991 from *Play-boy* magazine. We spent around $400,000 on that video, which had a ton of hot babes in it. That was the peak of MTV and the music video form and people spending money on them.

When I first saw the treatment to the video for "Right Now," the next single from the album, I thought it looked like a terrible idea. That was the first serious lyric I had written for Van Halen, a big statement. All I could see was some of the director's lines, like "I'll wrestle you for food" or "Right now someone's walking on a nude beach for the first time." I read his treatment and, without seeing the video, thought he was nuts. This song was my baby. I went around in circles with Eddie and Al for six months on this. It was the last song we recorded for the album.

I had these lyrics, but Eddie couldn't relate to what I was trying to say and he didn't have any music that worked. I had been pumping these guys for months while we were playing video games or eating—"Right now, it's your tomorrow . . . Right now, it's everything"—but nobody was getting it. One day during a break, I heard Eddie fooling around on the piano in the other room. I went running in.

"That's it, that's it," I said.

"I played this for you on the last album and you didn't dig it," said Eddie.

It fit like a glove. He didn't have to rewrite anything. If Eddie had gone off and played video games instead, or if I'd sat down and started making phone calls like I usually did on breaks, it would have never happened.

I felt protective of the song. I did not want to do the dumb video. The Warner Bros. brass tracked me down in Hawaii. They were trying hard to convince me. I told them okay, but let me and Al come up with the script. This director had scenes in his treatment like me looking in a mirror with a poster of the old Van Halen in it, while it says, "Right now, Dave wishes he was Sammy." We started writing together and began coming up with some pretty good lines. I still wasn't convinced, but I started to feel enthusiasm stirring.

When it came time to shoot, however, we were in Chicago in the dead of winter, stuck inside a blizzard. I had been ill with pneumonia and forced to cancel two shows. I had been trapped in a hotel room for days, sick, in a bad mood and pissed off. The director was so vague with us—"Just stand over here"—and I couldn't see the point of anything he was doing. At the end, there's a shot in there where I'm folding my arms, standing there looking disgusted. And that's exactly what I was doing. I wouldn't even sing. I was just throwing my arms in the air and he shot it and put it in the video. At the end of the day, I had a 102 temperature and was dying. I walked out of the warehouse where we were shooting into the bathroom we used as a dressing room. The guy with the camera followed me all the way. When I reached the door, I turned around and gave the cameraman a dirty look and slammed the door in his face and it goes MEN on the door. That's the way the video ends.

Mark Fenske, the director, what a great artist. It was the biggest video of our career, one of the biggest MTV videos of all time, and Crystal Pepsi paid us 2 million bucks to use it as a soda pop commercial. I can't believe all those people had to beat me up so bad before I caved in, but the treatment really was bad. The video was brilliant.

MEANWHILE, BETSY WAS back home, giddy happy. She was getting better. They put her on drugs and she snapped out of it, started doing real good. She had lost a lot of weight—Betsy wasn't hugely overweight, but, like any woman who's had a couple of kids, she had to struggle with her weight. The drugs slimmed her down. I was ready to start taking them. I was running every day on the beach or riding my bicycle into Santa Monica and back. She had her roses and her horses. Betsy was a Beatrix Potter kind of girl. She started taking tennis lessons. I should have been happy, but, instead, I was going, in my head, "I am damn near over this."

She had worn me out. I was done with that marriage. I wasn't going to leave her, I was going to make sure that she was okay, but I was over it. In that one year, she'd worn me out. Every night I had to hold her and rock her to sleep in bed. I had to feed her. I had to make sure she ate. To me, it was like having an invalid child. When she was on her medication, she was doing great, riding her horses, playing tennis, walking on the beach, and just being a normal person, like she had never been since I met her. I was glad for her, but she had been holding me down so hard for so long.

She didn't remember who she was. During that year, she was all broken down, really in trouble, in and out of mental hospitals. She forgot who she was. When she started taking Prozac and came out of her trance, she came back and was this happy person living in this beautiful house in Malibu with nice cars, horses, and money to burn. She flat didn't remember anything. She became like a shopaholic. She had clothes in the closet, expensive designer sweaters and gowns, things that she wouldn't wear in a million years, things that she just bought and shelved. She put something like a million dollars on her credit card that year. It didn't even register.

Once she'd leveled out, Betsy went back to wanting me to quit the business, especially since my other businesses were continuing to do well. She needed something to distract her. I told her to go find a realtor and look at property. She had a $2 million budget. I knew that would keep her busy.

She ended up buying a house in Carmel by Big Sur. All Betsy ever had wanted in life was to live in Big Sur. One time, maybe ten years earlier, we were at the Highlands Inn in Carmel and I was jogging through the neighborhood. I ran past the coolest house I'd ever seen. Right on the cliff, waves splashing into the windows, it was a Frank Lloyd Wright–type house that looked like an upside-down boat. It had this big copper roof with this big spine on the top, like the keel of a boat. I took Betsy to see it and she didn't even notice it, because of some storybook castle across the street. It was just as well. I had never dug it there—too big-spacey-heavy-lonely for me. We'd go down there from time to time and spend a week hiking and sitting there looking at the ocean, but it never felt like a place I could be in regularly.

Anyway, ten years later, Betsy was on her house search, when she called to say she found a place. We drove down and it was the same house I tried to show her ten years before. We bought it and Betsy threw a fortune in the place, totally doing it up.

Unfortunately, it didn't change things much. She was busy, sure, but the situation between us was as strained as ever. Our relationship was over, but I didn't feel I could leave her, because of the history. I'd been in this position before and wanted to leave Betsy a couple of times—not out of cruelty, out of necessity. It was like, "I can't live like this anymore." So many people go through that and stay together. They just learn to live separately for the rest of their lives, in separate bedrooms or whatever. We were still sexually active, so we weren't sleeping in separate bedrooms. But the only thing that was left in our relationship was sex, our kids, and an empire. We had houses, businesses, and cars. The thought of divorce was ugly. I'd never have gone through it if I'd known how ugly it really was. It wasn't just the unpleasantness of divorce, though; I just didn't want to leave her. I still cared about her.

I'm the kind of person who can put my head in the sand pretty good, as well as put my head in the clouds as good as anyone, too. I'm really optimistic about everything I do in life. I don't believe there's a downside to anything when I go into things. I've been let down a few times, believe me, but it doesn't destroy my optimism. I figured things were just going to get better. I was so busy all the time in my brain, it barely mattered.

Once I knew that she was okay and that she could be by herself, I decided I was willing to stick it out. I'd go back out on tour as much as I could and just keep fucking around and get out of the house whenever possible. I envisioned her getting better, maybe, but not our relationship. I didn't think that was ever going to come back around. It wasn't even on the top of my list. I was worn out from that year of taking care of her. I wasn't going to put more effort into it now, or try to make this relationship happen, like I'd done a year earlier when I'd told the guys I wasn't going to tour. I wasn't going to do that no more. I was optimistic that she would get better, but meanwhile I was just going to do my thing. I accepted that this was going to be kind of fucked up.

I went out on the For Unlawful Carnal Knowledge Tour, and I *really* started to mess around. I wasn't off the leash completely, and I was very careful not to get caught, very careful not to bring home some infection. I wasn't reckless. I was very, very cautious. I cared about my wife. It wasn't like I didn't love her. It's just that we'd grown apart. Everything else was the same. I still loved her family. I still loved our family. I even loved our lifestyle. She just couldn't stand my career and I loved it. She was in love with horses and I couldn't have cared less. We had become entirely different people.

Once I got out there on that tour, I was partying a lot more than I ever had. I had always been pretty conservative with the partying, but once Betsy was getting better, all that changed. I'd call home at night after she was dialed in on Prozac. I'd say I was going to dinner and might not call later. "Okay," she'd say, "then call me in the morning." Free pass.

I started having as much fun as a rock star could have on that tour. I had all the money in the world, all the babes a guy could want. I was having a really good time. Lead singer in the biggest rock band in the world and I took full advantage of it. I was eating in the greatest restaurants, drinking the finest wine, flying on private jets, walking onstage to sold-out audiences going crazy for us. The only thing missing was . . . I don't think anything was missing.

I fucked everything that walked. I had my own little tent underneath the stage. Eddie had his tent. Al had one, Mike had one. We all had our little tents. Mike and Al were on the other side. Eddie and I were on the same side, because Eddie was a dirty dog, like me. I sent roadies into the crowd to bring back girls I pointed out. During Eddie's guitar solo, which was always about twenty minutes, I'd have five or six girls in my tent, naked, all of us, having brutal sex while Eddie was out there doing his thing. When I went back out, I had to stuff my hard-on back in those tight pants. I'd wear my robe for the next couple of songs. That was every night.

That way I could sort out the good ones. After the show, we'd carry on. I had so much sex that it got to where I couldn't come. I'd go for two or three weeks without coming. It was like I was empty. I could fuck five girls all night. Some of the best sex I ever had in my life. John Kalodner loved this. He'd come out on the road with us and he'd line them up in his room. I'd go up there like a machine. "I've just never seen anything like this," he said.

And then Kari entered the picture. I met her at the end of the For Unlawful Carnal Knowledge Tour in October 1991. Every night I had been with a different woman. Our tour manager Scotty Ross was celebrating his birthday in Richmond, Virginia. Leffler liked him a lot and threw Scotty a party in his hotel suite.

Kari and two of her girlfriends came to the party. They were invited because her boyfriend was Buffalo Bills quarterback Jim Kelly, who knew the concert promoter. I was trying to pick up one of her girlfriends, because she seemed a little more available. Kari looked like a fun girl, really cool, good-looking, but I was chasing pussy.

Betsy was home, on medication. She was fixing up the house in Carmel, her dream home practically on the cliffs of Big Sur. I didn't even want to come home.

When Kari said she had to leave to go judge a beauty contest at some nightclub, Eddie and I, both trashed, volunteered to go with her. We went back later to Ed Leffler's party. I was good and hammered, still trying to pick up on Kari's girlfriend, when something we were eating fell on the floor. I looked down and saw Kari's feet. They're like fingers—really bony and, I was thinking, gorgeous, the most long, beautiful toes I'd ever seen.

I looked back up. I couldn't help myself. "You really have beautiful feet," I said.

When I looked up and saw her face and her eyes, I realized how beautiful she was. "You like those cheetos?" her girlfriend said.

She called her toes "cheetos." "Yes, I do," I said. "I'd eat them fuckers right now."

I started trying to hit on Kari. Everything was cool. We were having a good time just talking. At the end of the evening, around two o'clock in the morning, I invited her to come to my room.

"Oh, no," she said. "We'll walk you to your room. You're trashed. You need to get some rest. You have a show tomorrow."

She and her two girlfriends walked me to my room. I opened the door and she gave me a little hug. "Damn," I thought, "I've wasted all this time and I'm going to bed by myself?" I asked her if she wanted to go to the show the next night.

"I can't," she said. "I've got to take my grandmother to a wedding."

The lead singer of Van Halen, headlining the Hampton Coliseum, inviting her to the show and she's not going? She's going to take her grandmother to a wedding? I dug it. She got me, right then and there.

"What time are you leaving?" she said. "I'll try to make it back."

I gave her my hotel alias. She called the next night. "We're hauling ass in the car," she said. "We're going to try to make it. If we don't, put our name down on the list. We're coming."

They arrived just in time and Kari was wearing a bridesmaid's dress. They went into the other room and started ripping off their clothes. I peeked. They weren't getting naked, but stripping down to bra and underpants and throwing on some jeans. Kari, to me, looked like one of the most beautiful women I had ever seen in my entire life. And, I'm thinking, tonight's the night.

We were backstage at the gig. Scotty Ross pushed open the dressing-room door, still feeling the effects of his birthday party. "Scotty, this is Kari," I said.

"Nice to meet you," he said and puked all over the floor. Here I was, on my best rock-star behavior, and this guy comes in and blows it all over my dressing-room floor.

She was okay with it. On the way back from the show in the back of the limo, I tried to kiss her. She gave me little pecks. I had my arm around her and, for a second, I looked at her knees. Kari has these long, beautiful legs, skinny fingers and arms and toes. She's a slim, beautiful, smooth-skinned woman. Her knees looked like they were made out of porcelain. I leaned down and kissed her knee.

I put the goose bumps on her. She got them and I saw them. "That was sweet," she said.

Again, I asked her back to my room. "No, I can't stay over," she said, "but I'd love to see you again."

Whoa. Tonight's not the night? I planted a big lip-lock on her. We exchanged phone numbers, and, the next day, I left town with the band. About four days later, I called her. "If I send a plane for you, will you come and see me on my birthday?" I said.

"Are you kidding me?" she said. "Of course I would." She's got to give me some pussy now. She ain't got nowhere to go. I sent a jet to pick her up. It was kind of a last-minute deal. By the time she was delivered to the gig, I was already on the stage. I saw her on the side of the stage and my fucking heart just started flipping and fluttering. *Pow.* I fucking fell in love. I saw her and it was just like, shazam, there she is, my dream woman.

I am fucking anything and everything, four or five times a night, and, all of a sudden, I bit the bait. I swallowed it. I'm in the boat now, floundering around. I came off the stage while Eddie did his solo and, instead of having four or five chicks waiting for me, I'm sitting there holding hands with Kari like some schoolboy.

After the show, in the dressing room, the promoter and Leffler had a giant cake for my birthday, and a stripper came bursting out of it naked. Not just any stripper, either, but the fattest, ugliest, most cellulite-ridden babe they could find. There's no explaining how a girl that looked like that could do this for a living. I sat there, looking at Kari, trying to be this high-class guy, and thinking this was blowing it. But she cracked up. I thought, "I love this girl."

I told her she was not going home and she agreed to stay a few days. I had my Red Rocker tandem bicycle shipped out. I couldn't be away from her.

When she went back, I tried to spend a week without her, to test myself. I never fucked another girl. I didn't get a blow job. I didn't do anything, because all I wanted was her. I couldn't take it. And we were about to go home for the holidays after almost four months out. That was really rough.

She came and spent another week on the road. I was beat up. I did not want to go home. After she spent the week with me, that was it. I was done. I was in love. I was going to go home and give it one more shot with my wife. I arrived home about two weeks before Christmas 1991. Betsy had the house in Carmel all fixed up. I stayed three or four days, woke up in the middle of the night, and told Betsy I had to leave. She was on medication. She told me it was okay, to do whatever I wanted to do—this from Betsy, the most jealous woman in the world. This was when I wrote "Amnesty Is Granted."

I thought I could finally leave Betsy, because she was on so much medication that she was able, at last, to handle it. I was never sure before. One of the reasons why I never left her earlier was that I worried she was suicidal. I didn't want her to commit suicide because of me. I couldn't live with that. Betsy is not like other women. She does not fully belong in this world. She is harmless, vulnerable, and sensitive, but it is difficult for her to function. She can't be around pesticides. She can't eat certain foods. She is very intelligent and talented, but she is not strong in any way.

Finally, one night in Carmel, I confronted Betsy. I broke down. I started crying. I told her I was going up to Mill Valley to get my head straight and figure out some stuff. She was so zonked on her tranquilizers and mood elevators that she treated it as an unremarkable event.

"It's okay, honey. You just need some time. Whenever you're ready, come on back. If you have a girlfriend and you want to move back in the house with her, I'm okay with it. I don't care what you've done. It's okay."

I didn't tell her I had a girlfriend, that I'd fallen in love, that I'd been with seventy-five girls a week, practically, and suddenly I fell in love. I didn't know how to deal with it. I hadn't been in love in a long time.

I got in my car and drove home to Mill Valley. I called Kari and told her to come out and see me. "I just left my wife," I said.

"Oh, my God, that's terrible," she said, "but I'm with my grandma and my mom and dad. We're starting our Christmas ritual. I don't really think I can do this."

She finally agreed and I sent her a plane ticket. I picked her up at the airport and we went up to the house in Mill Valley. I felt so fucking uncomfortable. It was right before Christmas, and leaving five-year-old Andrew was breaking my heart. Then Betsy called. She had decided we needed to spend Christmas together.

"We've got a Christmas tree in the back of the truck," she said. "Andrew and I are going to come up. We've got a Christmas tree and a turkey and we're driving up to Mill Valley."

Betsy had this truck that I had made for her, an old '53 Chevy pickup on top of a brand-new Chevy drive train. Betsy, the horse girl, loved the truck. It was about a two-hour drive for her. She had not understood a word I said about leaving her.

I couldn't be there when Betsy and Andrew arrived. Kari and I jumped in the car and headed straight for the airport. First, we flew back to see her parents, where she had to apologize to her grandparents for missing Christmas. "If I'm going to do this, you're going to meet my parents," she said. "You're going to look my grandmother in the eye and say, 'I want your girl to come with me. I'm sorry. This is the first Christmas in her life she's not going to spend with you.'"

We flew back to Richmond, Virginia, and I met her parents. Her stepdad was pissed off ("That son of a bitch," I heard him say from the other room, "who does he think he is?"). We became best buddies later, but he didn't like me running away with his daughter. We took off for the Virgin Islands, a French resort called La Samanna. I was hiding out. Only Leffler knew where I was.

We fell in love. We stayed a month. I kept calling Betsy every few days. She did not understand that I was gone for good. She would tell me to take my time and come home when I was ready. As far as I could tell, Betsy never got her mind back together. She's very bright and sensitive, really human, and a great mother. But I sensed there was a screw loose that just wouldn't tighten back down.

When I finally told her brother, Bucky, his first words were, "I don't know how you did it that long."

9

RIGHT HERE, RIGHT NOW

The For Unlawful Carnal Knowledge Tour had been our biggest ever. We sold out so fast. We were doing two and three nights in the amphitheaters, a growing end of the rock concert field in the eighties, these huge holes scooped out of suburban earth that held twice as many people as the indoor arenas (universally known in the business as "sheds").

At the end of the For Unlawful Carnal Knowledge Tour, we'd decided to do the live album, *Right Here, Right Now,* if only to get a record out quick. We recorded and videoed *Right Here, Right Now* in Fresno as the tour was ending. After that Christmas with Kari, she and I had taken off on a rocket ship. We flew to Maui and stayed there for three months, and while Kari and I were off on our rocket ship, the Van Halen brothers were supposed to be in the studio mixing the live record. It should have been simple, but the dumb-ass brothers decided to take the live album, because they were so bored, back in the studio.

That's when we started bumping heads. Looking back, I can see what happened. Al's marriage was washed up. Eddie's marriage had been on the rocks for some time, or at least that's how Valerie tells it. It got to the point where Eddie was pretty wasted much of the time. Eddie went to the Betty Ford Clinic. He did rehab a few times. It never lasted more than a couple of weeks. When Al quit drinking, nobody changed. Eddie was drinking in front of him. But when Eddie came out of rehab, suddenly the rule was no booze in the studio, not that I ever sat around drinking beers. Sometimes after we'd finish recording, I'd bring in a bottle of tequila and Mike and I would do a couple shots, laugh, and have a good time. And, of course, Ed would do them, too. He wasn't sober. He would keep everything stashed in the studio, with Valerie at the house next door.

He never wanted me to go home. "Why do you have to go home now?" he would say. "Wait a minute, I got one more thing. I don't like that part. We've got to recut that. You've got to re-sing this thing here." He wanted to keep me as late as he possibly could, because he didn't want to go home. Because once he went home, he couldn't get back out and get to his stash. And once my car left their driveway and Valerie saw the lights and heard the engine and the gate open and close, then she knew I was gone. It got to the point where he was getting all fucked up late at night and making stupid remarks, because he wasn't on top of it. There was always some reason. He would never cop to that he wanted me to stay because he didn't want to go home. Valerie's home, he's in the studio. Valerie leaves, he's in the house, drinking.

As the mess got bigger, our conflicts grew. Eddie Van Halen, who had been the humble guy under his big brother's thumb, wanted to take over his band. He'd always been kind of passive-aggressive, but it got difficult to deal with. He would be humble and back down from confrontations, but then he would go behind my back and complain to Leffler that I wasn't working hard enough.

And that was what happened when Kari and I went to Hawaii.

The Van Halen brothers started freaking out, getting down on me big-time because I wasn't around to rehearse the new studio sessions with them. I'd left my wife and now I didn't want to go and rehearse. Fuck that.

The problem was they'd rerecorded almost the entire live album. Because Eddie was out of tune, or Al had sped up or slowed down, they're fixing things. They fixed everything. Only now that Eddie was playing in tune, my singing's off-key. And where Al sped up in "Runaround," now I'm singing ahead of the beat. Now I had to go back in the studio and redo all my vocals. I wanted to kill those guys.

Kari and I flew back to Los Angeles from Hawaii. I told Eddie to stay the hell out of there. They put me in a room with the video of the concert, gave me my microphone, and I stood there and sang the whole fucking concert one time through. Just like it was a live performance. I barely went back to fix anything. It took me three hours and then I went to dinner.

The brothers were pissed. They took out the microscope, trying to find places that weren't reasonable, that I needed to fix again. When they found something, I went out and fixed it. Fuck you.

Meanwhile, as Kari and I had been grooving in foreign ports of call, Betsy had filed for divorce. I hadn't gone home since I left at Christmas. It broke my heart to leave five-year-old Andrew behind, and it was going to be a few years before I saw him at all. The split was not going to be amicable. It was going to be difficult and expensive. Betsy hired a lawyer who was picking over everything. He wanted to pay a recording engineer to go through and catalog all the unrecorded music and song ideas on cassettes I had at home, in case I wrote any songs in the future based on material I started when we were still married. I tried to head off everything with a settlement offer that would have actually given her more money than she got three years and millions of dollars in attorneys' fees later.

Leffler figured out a way I could pay for the entire settlement in one swoop. He told me Geffen would pay huge money for a greatest-hits album—I'd never done one—and if I came up with a couple of new songs, there would be generous publishing advances for each of those. He worked it out so that the one album with the two new songs would pay for the entire divorce.

I went to see my attorney, who had drawn up the deal, and sitting in his office, next to each other, like they do when they're insecure, were Eddie and Al. They didn't want me to do the album. I told them I was going to do it and that it was going to pay for my divorce. They argued and argued. They said it would be bad for the band. It wasn't like they said anything before to my face. They were real behind-the-back guys. They had a couple of conversations, worked each other up, started freaking out, and began to look into ways to keep me from doing something I wanted to do. Eddie didn't know anything about his business. He probably didn't even know where his money was. When my lawyer mentioned something about my getting paid on a publishing deal, Eddie wanted to know how he could get one. I told him he already had a publishing deal. They just didn't want me to do anything they couldn't control. It wasn't long before they started seeing attorneys of their own about possibly suing me or throwing me out of the band.

SINCE *RIGHT HERE, RIGHT NOW* was going to be a double-record album, Warner Bros. raised the price. That hurt sales. It went to number five on the charts, our first album not to go number one. In the summer of 1993, we went out on another huge tour to promote the live album, and we made tons of money on it.

Even though the tour was big, the Van Halen brothers were still working that attitude on me that I wasn't doing enough. "If you could sing five nights a week, think how much money we'd make." They didn't care about my voice. "If you can't sing, just dance," they'd say.

We started getting into it more often, and things weren't as friendly. Eventually, I started flying on my own. I'd fly home by myself, and I'd come back. I'd stay in different hotels. By the end of the tour, Eddie and I weren't getting along.

What complicated things was the fact that Ed Leffler got really sick in the middle of that tour. At the start of the tour, he'd found a lump in his throat and it was cancer. He had it taken out and came right back out on the road. He even stopped smoking for a while, although that didn't last. But you could tell he wasn't doing so well. He was always sweating, kind of pale, and losing weight.

The cancer came back. He hit it with chemotherapy and radiation, and it spread. He was done. It was only a few weeks. He went from a guy out on tour with us, getting pussy, doing blow, drinking, having a good time, in his mid-fifties, and now he was going to die. He was weak and sickly, but stayed on the road with us.

On the last two nights of the tour in August 1993, we were playing at an amphitheater in Costa Mesa, outside Los Angeles. I was feeling blue about Leffler, so I decided to switch the acoustic number I usually did, "Eagles Fly," to "Amnesty." Since we were back in town, all Eddie's bad-news friends showed up with the drugs and the women, and he was wasted. In the middle of my song, he decided he needed to change the tubes in his amplifier. I'm out there doing this song and Eddie's over there panicking, taking his equipment down behind me. I'm trying to do this sensitive song, and it was really pissing me off. I'm playing acoustic and singing "Amnesty Is Granted" and Eddie's checking his tone out to see if the tubes were working, fucked up out of his mind.

I came off the stage and grabbed him. We got into it, but Leffler pulled us apart. I came out for the encore, waiting for Eddie. I was going to kick his ass right here, right now. Leffler shoved me in the back of a car and off it went. Later I got a call from Al telling me that not only was that the first time we didn't do an encore, but Ed Leffler had collapsed. His legs went numb on him and he fell down and couldn't stand. They had him lying down.

Eddie apologized and I came back the next night. He was like that. He would do the worst shit you could ever imagine, and the next day he'd be humble and whiny, crying and hugging you. It was easy to forgive this guy, because he went all the way to the ground with his humility. Next day? Whole different guy.

The next night, the final show on the tour and our second night at Costa Mesa, we did one of the greatest shows we'd ever done. We were worn out, beyond tired. It was the end of the tour. We went out there and played from a whole other place for the first time in a long time. We played a real emotional show. Every song felt like everybody meant it. We weren't just doing a regular show. We burned our encores and everything down to the ground. The next day, Ed Leffler went into the hospital.

We did all this crazy shit to try and help Leffler stay alive. I found a lady who took urine specimens to Mexico, where they took the neurotransmitters out of your pee and returned it in little vials. You shot it up every day, in the muscle, not intravenously. I made Ed do it. I did it with him. And then there was this purple goo, slime that dyed your skin violet. We put his feet in it, supposedly to suck all the toxins out. We tried everything. He was walking around, breathing oxygen out of a tank.

We kept him alive for a month, maybe, with all these different things. He didn't have any hope. Leffler would just look at you and say, "Sure, okay, I'll try it." He was a real smart guy, but he didn't believe in hocus-pocus.

By October, I was getting ready to go to Mexico with Michael Anthony for my birthday. I went to see Leffler in the hospital the day before. He was in bad shape. They were taking a liter of fluid a day out of his lungs. He was on morphine. He asked me to massage his hands. He couldn't feel anything. I'm massaging Leffler's hands. I ask him who should we get to manage the band. "Just stay away from Howard Kaufman," he said.

I couldn't believe it. Howard Kaufman managed the Seattle rock band with the two Wilson sisters, Heart, and Leffler was holding a grudge from a while back when Kaufman had pulled Heart and all the other bands he managed out of my travel agency because he thought Leffler owned it. Leffler didn't care who managed us after he died, but he took his enemies to the grave.

A couple of days later, Eddie and Al called me in Mexico to say that if I want to see Leffler alive again, I'd better get right back. I had a big celebration planned for my birthday and didn't want to leave. I called the hospital and talked to Leffler, who told me everything was fine, to stay where I was.

The next night, I was going with my brother to play at the cantina, and I felt a cold wind blow through me. I looked at my brother.

"Wow, I just got the loneliest feeling—I feel lonelier than I've ever felt in my life," I said.

I didn't think Leffler had died. I wasn't even thinking about Ed Leffler. Honestly, I was thinking about the gig. I walked out in this beautiful, warm Cabo night, and something walked right through me. I felt like the only person on the planet.

I got the call at the club. Leffler had died. Just like with my dad. My brother was there—he saw it happen. I did the show and got on a plane the next morning. I went to the funeral, and did a little speech for Leffler. When he died, they put a gram of blow and a bottle of J&B Scotch in his coffin. His friends were characters. They didn't take it lightly or unlovingly, but they did this crazy stuff. That was the end of Ed Leffler.

❊ 10 ❊

CABO WABO

In December 1983, I saw a photograph in *People* magazine from the wedding of Keith Richards and Patti Hansen. They were standing poolside at the Twin Dolphin, the only real hotel in Cabo San Lucas, Mexico, at the time, and I thought it looked cool. He has always been one of my heroes, and I told Betsy we should go down, check the place out.

There was only one flight a week—one flight in, one flight out—and two places to stay with dirt roads from the airport all the way to the Twin Dolphin. There were no telephones, no newspapers, no televisions, and no air-conditioning. To make a phone call, you had to go to the phone company downtown and pay by the minute after they placed the call for you.

When Keith came down for his wedding, he'd planned to stay for a week, but didn't leave for three months. His family went back after a couple of weeks, but he stayed, sleeping on people's floors. Jorge Viaña, the bellman at the Twin Dolphin who eventually became the manager of the Cabo Wabo, took Keith everywhere. Keith liked to sit in with the mariachi bands. They didn't know who he was, this crazy gringo in the rock-and-roll clothes, but he was drinking tequila straight from the bottle and handing out $100 bills, so they loved him.

Keith borrowed Jorge's car to run into town and make a phone call, about seven miles into town from the Twin Dolphin. He never came back. Nobody had cars down there, but Jorge finally talked the manager into driving him downtown after midnight to look for Keith and his car. He saw his car parked in the service station, long-closed. He looked inside and there was Keith, passed out on the floor next to the service station guy, a couple of empty tequila bottles beside them. He probably just stopped to get some gas, and maybe some directions.

Betsy and I went down there shortly after I read about Keith's wedding, and I fell in love. It was such a pristine, beautiful place. You would be walking on the beach and a wave would crash on the shore and blast a five-pound red snapper up on the sand. All you had to do was reach down and pick it up. You could snorkel anywhere in the rocks and pull up oysters. You could damn near catch fish with your hands. There was nobody around for miles.

The place was pretty much closed up during the summer, and if you didn't want to eat the mediocre food at one of the hotels, your best bet was a local taco stand or somebody's house. Latinos are very hospitable about inviting strangers to eat at their homes, even people they meet on the street. I used to go to people's houses and eat, all the time, in Cabo.

While I was down there, I'd run across this place outside of town, called Guadalajara, a little *palapa* shack, no windows, no doors. Chickens were running around. I sat down and, in my little bullshit Spanish—Betsy could speak Spanish pretty well but I was lost—I asked, "What do you have?"

"*Pollo, frijoles, arroz,* and *cerveza,* chips, salsa," the man said. Outside, walking across the street from the marina, he saw two kids, not more than eight years old, carrying a big swordfish on a stick poked through its eyeballs. Each kid had an end of the stick and they were dragging the giant fish. The guy turned around. "Y *pescado fresco,*" he said.

He goes over to the kids, gives them some money, takes a knife, and, *whap, whap,* cuts a couple of big steaks off the damn thing. They go off down the road to the next restaurant with the fish their pop had just caught. I ordered the fish. I thought I'd died and went to heaven. This was the coolest thing I'd ever done in my life—sitting there, having a brew, not a car anywhere, chickens running around. You're throwing crumbs down, the chickens eat them. You see a chicken on the grill, you know where that sucker came from.

About the third time I went, still before I joined Van Halen in 1985, Jorge took me to town. It was all dirt roads. You couldn't drive down there. You'd run out of gas in between gas stations. The first time I drove there, we had to sleep on the side of the road at a gas station, waiting for it to open. There was nothing there but little shacks for restaurants, but you could see the marina from anywhere downtown. I decided I wanted to build a bar. I had already tasted real tequila and fallen in love with the stuff. I told Jorge to find me a piece of property. I had a phone put in Jorge's house, so we could stay in touch. They ran the wire from the phone company office, wrapped it around stop signs, and took it to his place. Fortunately, he lived right downtown.

There was a new development, called Terrasol, going up on the most beautiful stretch of beach around. They only had one condo finished when I first saw it, but I bought a place on the spot and moved in for the whole summer. That time of year, Cabo was a ghost town. Everything was closed. You couldn't even find a restaurant open half the time. The condo units were all empty. I had a couple acres on the beach and a giant swimming pool to myself.

I also started going down every October, because my brother's birthday is October 8, my sister's is October 11, and mine is October 13. I took my mom, my brother and sisters and their families to the Twin Dolphin, and celebrated our birthdays for two weeks. Bucky shipped down one of my hot-rod mountain bikes and I was biking the dirt roads around Cabo every day.

One year, Jorge told me about a triathlon that was being held by the local military base but was open to outsiders. I signed up. The city offered a $1,000 cash prize, so the race attracted a lot of interest. Pretty much the whole town turned out at the marina to watch, as maybe 150 contestants crowded the dock where they were to swim the first lap, a quarter-mile across the bay. I was wearing a regulation banana hammock, a pair of Speedos, but everybody else simply stripped to their underwear there on the spot and jumped in. I held back. A lot of these people didn't know how to swim and they were splashing and floundering.

I dove in and swam across. When I got out, there was a short, stocky guy who started running down the beach, leaving me in the dust, but I was up there with the top four or five guys. We reached the third leg, the bicycle part, and Jorge was waiting, holding my bike. Only about half of the people still in the race, running down the beach behind us, had bikes. I didn't know what the others were going to do. The bikes they did have were trashed, big, heavy clunkers, some even missing tires, and I had this lightweight, ten-speed mountain bike. They had never even seen anything like it down there. I was beginning to feel like the biggest asshole in the world. The short, stocky guy was chugging up ahead of me on his junky piece of steel. He was dying going up the hill when I went flying by him in my mid-gears.

I finished so far in front, I was barely breathing hard when he finally made it across the finish line. I looked at Jorge and handed the heavyset dude the trophy and the cash. He took them both, turned around and raised his hands above his head like he won the race. He never even said thank-you. Jorge and I laughed at that for days.

The more time I spent down there, the better it got, and once I was in Van Halen, Cabo became an important part of how I wrote songs. I would jam new tunes with Ed, Al, and Mike and make up lyrics by scatting along. Then I'd go to Cabo. I'd relax on the beach, finish my lyrics, come back and do my vocals.

A couple of songs from *OU812* had actually come from my writing down there. "Sucker in a 3 Piece" came from Cabo. I saw this gorgeous chick poolside at the Twin Dolphin, who was married to this old dude, and this chick's giving me some vibes even though I was married and she was with this rich guy, the "Sucker in a 3 Piece."

One Sunday, about nine-thirty in the morning, I was driving to my favorite taco stand for breakfast, down a dirt road with a barbed-wire fence. Some guy was wobbling down the road in front of me. I couldn't get around him. He bounced, first, off the fence into the road in front of me, and then back into the barbed wire. He had blood running down his leg and was missing a shoe. He was some local who'd been up all night drinking mescal. I watched him make his way down the road like that, and it occurs to me— this guy is doing the Cabo Wabo. I went back to my pad and wrote the lyrics. "Been to Rome, Dallas, Texas, man, I thought I'd seen it all—round the world, every corner, man, I thought I'd hit the wall." The whole song spilled out of me, "Cabo Wabo."

Since I had one of the only telephones in Cabo, I called up Eddie and said, "Eddie, listen to this." I wrote the song, in my head, to the music from "Make It Last," one of the first songs I wrote in Montrose. I sang it to him over the phone.

"Oh, man, listen to this," he said. "Al and I worked this up last night." He played some music that sounded very much like "Make It Last."

On the phone, it worked. I flew back to L.A. early so I could sing that song. They had recorded the music while I was gone. I walked in. I took a handheld microphone. I was just going to scat, but I read the lyrics off my paper from beginning to end, and that song was done. On *OU812,* my vocals have a funny sound because of that bullshit little handheld mike, but it was such a perfect vocal take, we all decided to keep it.

IT TOOK HIM nearly four years, but Jorge finally found a piece of property for the bar I'd wanted to build, but it was going to be expensive. Even though I was in Van Halen, I didn't have the money for a million-dollar project, more like a half-million-dollar project. But still, that was a lot of money to be putting into a town with dirt roads and no telephones.

The town had been building up a little bit though. None of the side streets were paved, but they'd paved the road to town and partway through town. The swinging set in Hollywood began to discover the town. The hipsters were coming down. It wasn't just a little fishing village any longer. Boat-owners and private-plane pilots found the place. Private planes would land on this dirt strip. Walking around, you could sense the potential, it just wasn't there yet.

I knew I was going to call the cantina Cabo Wabo. I had already written the song. It was going to be a tequila bar, a small place with a stage. I told Jorge to find an architect, and he found Marco Monroy Jr., son of the developer of Terrasol, who I had met. His father had showed me a smelly, old sardine factory earlier, when I was looking for locations. His son had recently graduated college and was starting to work for his dad. He had built a couple of houses that were the coolest houses down there. I hired him.

Marco did the plans. It looked fantastic. I thought the building was three thousand square feet, but Marco and Jorge were talking three thousand meters. I thought there would be plenty of room for a big parking lot. When they laid the foundation, it was three times larger than I figured. I was thinking of a nice, small room that would hold 50 or 60 people, 150 tops.

With everything going on, the plans for my cantina had begun to get a lot of my enthusiasm and Eddie and Al couldn't help but notice. Finally, Ed Leffler had told me the other guys in Van Halen were beginning to feel like they were being left out. He'd taken me aside and gently suggested that I make the other guys my partners in the cantina.

"You want MTV to really support it, you want the press, you want the publicity," he'd said. "Bring these guys in on it."

We held a meeting and all of them, including Leffler, decided to join up as equal partners. They each gave me some $70,000 to repay the money I'd put up.

It worked. Van Halen played the gala grand opening weekend in April 1990. MTV spent millions of dollars on a big promotion. They filmed commercials and held contests. They flew a whole airplane full of people down there. Raquel Welch was there. Brad Delp from Boston, Steve Lukather from Toto. The whole town was excited.

Betsy's freak-out on the plane had been the year before and she hadn't been on a plane since. By the grand opening of the cantina, she was beginning to come around and the medication was starting to work, but she wouldn't get on a plane. I still had my plane, but I couldn't fly down there without her. I bought a motor home to drive us all down for the grand opening. I got my brother-in-law and my sister to come with me, Betsy, the nanny, and the kids. It took three days to drive down. It was my second time driving down and it was rough going. By the time I got there, I wanted to kill Betsy. In the middle of the drive, I sent my plane down to

Cabo with her psychiatrist, our doctor, and their families, and a couple of friends of mine. I put eight people on that sucker and flew them down in my plane while I'm driving a fucking thirty-two-foot motor home for three days.

Betsy loved Cabo, but she was afraid of everything. She was afraid that she was going to have another panic attack. She hadn't been back since she flipped out the last time. She was nervous about going back to the same condo in Terrasol. She was in kind of bad shape. We were on the outskirts of town, twenty minutes from the condo, showing our guests one of her favorite beaches, and she was getting disturbed. The psychiatrist suggested they go for a walk on the beach.

She was shaking and he was trying to calm her down. While they walked down the beach, we sat around, trying to give them some space. I snapped. I cracked. The pressure of the grand opening, and the Van Halen guys coming down for the first time. They were all freaked out, too—What, no telephones in the room? What do you mean, no room service? I stormed over to where Betsy and her psychiatrist were talking.

"Fuck this," I told the psychiatrist. "I've got shit to do. We're getting out of here." She actually snapped out of it. It woke her up. We got back to the condo, and I knew I couldn't be too heavy with her.

We left to do the sound check. Betsy's doctor came. The guy had never seen me play before. He was just a psychiatrist who knew Eddie. The doctor was blown away by the sound check. "I've never seen anything like that in my life," he said.

The grand opening weekend went great. Van Halen played a couple of nights. We had MTV and Mexican TV. It was a really big deal that went sour almost immediately.

The first week, everything went great. As soon as we left and the town emptied out, nobody came to the cantina. The locals didn't go to Cabo Wabo. We didn't have much of a restaurant, only a big taco bar. We served drinks. There really wasn't much to do there. We didn't have a live band. We played music over speakers. It was a fourteen-thousand-square-foot echo chamber. It was dark. We had a lot of low lights and everything was black. It really didn't have any charm yet. We built this place and opened it. Marco wasn't involved. Jorge, who never did anything like this before, was running the business. Once a week a plane would come in, and there would be people in town. Once or twice a week, the place would do well, but not that well. The rest of the week, it was empty.

T-shirts were selling well when we first opened. We could never get Jorge to send the T-shirt money. Jorge stopped because he didn't have any money to buy more T-shirts. He didn't have money to buy more booze, more food, pay the employees. The place was dying. It was losing about ten grand a month, which is plenty of money. Jorge didn't know what he was doing.

Leffler was still alive and healthy when all this was going down. He'd flown down to sort things out and on the plane he met the Deadhead son of the man who owned another hotel in another town. I knew him from the hotel. He sat at the bar, drinking all day. Leffler fired Jorge and put the Deadhead in charge. Didn't do a damn bit of good. He had more business sense, but the guy was doing drugs and drinking and the Federales were shaking him down because they knew he was dealing.

Van Halen only played Cabo one more time, after a Mexico City concert in 1992 on the cantina's second anniversary, but Mikey and I used to take David Lauser down to play my birthday bash in October every year. Eddie and Al weren't happy about the place. Every time they turned around, we were asking everybody for money. They each put in another ten grand a couple of times. That would support the place for six months or so. It was losing more than a hundred grand a year. After a couple of times, they said they weren't going to put any more money in the cantina.

Finally, Mike and I decided that every time the cantina needed money, we'd go down and play. We'd do two or three nights, the place would be packed. We'd charge five bucks at the door. That way we kept it going. We played the cantina five times that year and never had to put money in it again.

One night, our Deadhead manager asked if he could introduce me. He seemed so coked up, his jaw was going from left to right, grinding his teeth. He got up there and started telling jokes and stories. People were throwing stuff at him and yelling. We had to drag him off.

In the office, I told him to open the safe and he was so addled he couldn't work the combination. He finally opened up the safe and the only thing in there was a bag of coke. I fired him. I went back and told Leffler. The band was all pissed off. It was a mess.

He later cleaned up and told everybody he was sorry. He had a family and lived down there. He was trying to get it back together. Leffler cut him some slack, because we didn't have any choice, but the whole situation had made all of us—especially the Van Halen brothers—anxious about Cabo and where it was headed. They refused to sink any more money into it and seemed like they were done with the whole idea of the place. I thought we could keep it going, but I knew it needed help. What I didn't know was that after Leffler's death, things would only get worse.

FATHER'S DAY

When Leffler died, we auditioned managers. I wanted Shep Gordon and Johnny Barbis—Shep was Alice Cooper's brilliant manager and Barbis was one of the best-liked people in the business, ran labels, was pals with U2, Elton John, everybody. The brass at Warner Bros. liked the idea. We met with them. The brothers didn't like them. I called David Geffen and he suggested his old partner, Elliot Roberts, Neil Young's manager. We met with him, too, and the Van Halens blew him out in about five seconds.

Ed and Al wound me up for two months auditioning people before they told me they wanted Ray Danniels. He was married to Al's wife's sister and managed Rush. They told me I got my man the last time and they wanted their guy this time. Ray Danniels had been lurking in the background the whole time.

Before he was even our manager, Ray Danniels had told the Van Halen brothers about a publishing deal Leffler made on the live album that I didn't know anything about. It was no big deal, but Ray Danniels made the brothers think they'd been screwed. They made me pay them back a substantial sum. Alex Van Halen never wrote a song in his life and he was taking the same amount of publishing money as me. Danniels gained the confidence of the brothers that he was going to be on their side, not mine. Ed and Al were really going against me at this stage. They thought Leffler and I fucked them. We didn't fuck those guys. We saved them. They made ten times more money in one year than they'd ever made in any year before we came into the band. Nobody fucked anybody.

I told Ray to his face, "They're going to sign with you. I'm not. You get zero of my money." The deal I made was they paid management. I didn't pay management. He didn't do shit for me. He wasn't my manager. I would find my own manager. I did not like the guy. I wanted to bite his face off. And he was scared of me. He didn't want to come into a room with me. He stayed away from me, always holding meetings with Eddie.

Ray Danniels went to Warner Bros. and renewed our contract. He negotiated a few extra points for the band's early albums—the ones I wasn't on—other than that, nothing changed. He renegotiated the same deal we had to begin with. Except for one thing. When I joined Van Halen, Ed Leffler had put in the contract that after every Van Halen record, I had the right to do a Sammy Hagar solo record for big money. I only did the one. Ed Leffler called it my golden parachute. Somehow they took that out.

I walked into a dressing room backstage in Toronto, Ray Danniels was there with his briefcase. Ed and Al were signing papers with a notary. They were signing the record deal and they didn't even want my signature on it. "Don't worry about it," Ray Danniels said. "Ed and Al are all that matters in this band."

A FEW MONTHS after Leffler died, the cokehead manager down in Cabo called in early 1994 to tell me he gave the keys to the employees and that the government had wrapped a yellow ribbon around the place and closed it down. I couldn't think what else to do, so I called Marco Monroy.

Marco discovered the manager didn't pay any bills for the whole year. He spent all the money. Marco said the cantina owed around $300,000. The place was trashed. The furniture was shot, the equipment broken. He wanted to be my partner. He offered to pay the debts and invest another hundred grand into fixing the place up.

Jorge, by now, was long gone. He hooked up with an American "actress" with some bad habits. The problem was that everything was in Jorge's name. He was gone. We didn't know where the hell he was. It wasn't pretty.

I needed to take complete control of the cantina. The Van Halens had already told me to shove the place up my ass, and after Leffler died, our relationship got even worse. I went to Ray Danniels and asked him if I could buy out the other partners. He was trying to get on my good side. He cooked up a plan with our accountant where they could get their money back by taking the loss. They wrote it off on their taxes. They gave it back to me after I agreed that if I ever built another one, they would have the first right of refusal to invest. If I sold it within five years, they would get their investment back, although that would be a little tricky after they already took it off their taxes. I had to indemnify them against the debts and any other legal problems. It was a little complex, but I went for it.

Marco wanted to bring in someone he knew to manage the cantina. Tito was a tough *hombre,* married to a wealthy Mexican heiress. They lived in a mansion Marco built. Tito cleaned house. He not only tightened up the staff, he got rid of the drug dealers and lowlife's who were hanging out at the place. Marco and I decided to get the title to the property back and get the business into shape. When we couldn't find Jorge, we went to his ex-wife, who still lived with her kids in Cabo. We offered her $25,000. She didn't speak English, but we brought an interpreter. She got up and walked out. I have no idea why, because she ended up with nothing.

We finally dug up Jorge and he hard-nosed us. He wanted 10 percent. Marco and I each gave up 5 percent to get him to sign off on everything. Before long, he came crawling back asking for his job. He left the chick. He was straightened out. He moved back to Cabo, and we let him come in. He's been there ever since and he's been, as much as Marco, a savior of the place.

The town was all starting to come together. The big dream was really happening in Cabo. The road was paved now. There were more hotels. Three or four planes were coming in a day. The town was packed. From the first day, Marco and Tito really turned Cabo Wabo around. They cleaned the place up, made it nice. Boom, within the first month, we started making money.

The place looked fantastic. We were putting money back into it and taking money out without any out of our pockets. The first year we made around two hundred grand in profit.

The brothers weren't happy. They started accusing me of running the place into the ground so that they'd give it back to me. I wish I were that smart. Scotty Ross, our tour manager, sort of a big-mouth guy, came back from Cabo and walked into a Van Halen rehearsal and slapped my hand. "Cabo Wabo was packed, dude," he said. "You're making tons of money. The place looks great." The Van Halens weren't smiling.

Mikey and I were still going down almost every other month. Mike was willing to roll with me. He was planning to go down with me for my birthday that year, but they wouldn't let Mike ever go again. Mike wasn't allowed to go to Cabo. They really thought I fucked them.

❋ ❋ ❋

AS CABO WAS coming together, I was spending as much time as I could with Kari. She and I just wanted to go do things. We spent every night together. We lived in Hawaii, Mexico, Mill Valley. We'd go to New York. We'd go to Malta. We went to Italy. We went anywhere we wanted to go.

We had such synchronicity. We were walking around at the Hana Ranch on Maui, when the idea came up to get a parrot. We walked a little farther, and there's a cage with some parrots. This one little gal came to the cage and rubbed her head against the bars, like birds do. We cut a deal and took her back to our room. Her name was Spooch. That bird slept with us under the covers.

As soon as we got home to Mill Valley, we were sitting in the backyard by the pool. We clipped Spooch's wings so she couldn't leave. She was sitting on our shoulders. Spooch was a nanday conure and we were talking about how we had to get Spooch a boyfriend, when out of the sky flies a motherfucking nanday conure, who lands on Spooch's cage. Spooch was talking to the bird. The bird went into the cage to drink some water. Boom. We got us another bird. We called him Spooky. He never got along with Spooch—they fought like crazy—and we eventually had to give him to Bucky, but that happened to us.

Shortly after we got together, we were in Boca Raton, near where Kari's father lived in Florida. She wanted to go see her father. He bought houses, would move in for six months, fix them up, move out, and rent them. We took a limo—it was more than an hour away—and went out to dinner. In the limo after dinner, we smoked a fat one and started to get a little sexy in the backseat. We found her father's place around midnight, and Kari grabbed the key from the top of the water heater. We let ourselves in, turned on the living room light, and started rolling around on the couch. I put my foot on the floor and, hey, stepped on a pair of men's shoes. I jumped up naked and turned on some more lights. There's a shirt across a chair, an ashtray with cigarettes in it. Somebody has rented the house and is living in it.

We ran out of there, stoned on our asses, half-naked, throwing on our clothes, into the limo. Our hearts were pounding. We could have been killed, but we couldn't stop laughing.

So many things happened with Kari and me because we were in sync. With Betsy, I was living a lie. I was lying to her about everything, and I was therefore lying to other people on the phone, because she could hear. I was this whole lie—so far out of sync that things weren't working for me. As soon as I fell in love with Kari, I opened up and never lied again. I felt spontaneous. I felt free. Everything we did was the right thing to do. Things came to us that we wanted. You would think it, and it would happen.

We could laugh over anything. Another night, after we moved back to Mill Valley, at the Sweetwater, the town's tiny rock club, I met Bob Weir of the Grateful Dead. Kari and I ran across him sitting at a table with a girlfriend, drinking Suntory scotch straight out of the bottle. We sat and drank together until the place closed at two in the morning. "Let's go to my house," he said.

The smart thing to do is have everybody come to your house and then they have to drive home, not you. "No," I said. "Let's go to my house."

He pulled into my driveway in this beat-up old Corvette that hadn't been washed in years. My driveway is pretty wicked. There wasn't even a curb, just a sheer drop that goes down 250 feet. Bob brought a mason jar full of buds. His scotch was almost gone. We started smoking weed and continued drinking. We played a little guitar. We peed off the deck. About four in the morning, I'm wasted and shot. I told Bob it was time for them to go. We let them out the door and Kari asked wasn't I going to help him get out of the driveway. I didn't see it. He was Bob Weir of the Grateful Dead. He could take care of himself. I started up the stairs.

A terrible loud scraping noise outside—Kari's going, "Oh, my God"—and I come running back down. Instead of backing up and turning around, he has driven straight forward, off the driveway, and his car was now sitting half on and half off the driveway, facing down, the rear wheels off the ground. Kari and I dashed out and sat on the trunk. Weir, sitting in the car next to his girlfriend, looked dazed. "I think I need to pull forward," he said.

I told him to sit very still and have his girlfriend climb out carefully. She crawled out over the back and sat on the trunk with us. He climbed out after her. The car could have gone down in a second. No seat belt, convertible, down the hill—he's dead. Bob Weir dead at Sammy Hagar's home. It was horrible and five in the morning. I told him to walk home. He took his mason jar of buds and wandered off down the hill with his girlfriend.

The next morning, I had to meet some people and I was very hungover on about three hours' sleep. I backed out my Ferrari down the driveway, trying to maneuver around his car and not knock it down the hill, and broke off my goddamn $1,600 side mirror. I called a tow company. When we got home around five o'clock in the afternoon, the tow truck left a note saying they couldn't take the car. There wasn't enough room behind the car and they were worried about pushing it over the hill.

It took two tow trucks and three days to get that damn car off my driveway. I even paid for the tow truck. I was pissed at Bob, but Kari and I only laughed.

After a while, Kari started wanting to settle down a little more in the house in Mill Valley. She started putting some of her own items in there. She'd been living in Betsy's house, sleeping in

Betsy's bed. She started making changes. I began to see her domestic side. Before long, she started talking about how she would like to have a baby. Andrew was about ten. I was reluctant. But when a woman says she wants to have a baby, you don't tell her no. But I wasn't into being a father again.

Aaron was grown and living in Los Angeles, but Andrew was a heartbreaker. At first, Betsy wouldn't let me see him. She finally started letting him come up for weekends, but it was tough on the little guy. I'd catch him crying in his bedroom. The divorce was hard on Andrew. That's something I'll always carry with me. Being a father again didn't look all that attractive.

Then I remembered what Miss Kellerman had told me about moving to Northern California: "Someday you're going to have two daughters." I realized everything else Miss Kellerman said came true.

Sure enough. Just by bringing it up, Kari got pregnant. We were in the Jacuzzi, out by the pool, middle of the day, and we threw it down right on the grass. I knew it. We made wedding plans.

KARI AND I had been together almost four years. Leffler had been dead almost two years. The band had not toured since that last weekend in Costa Mesa. We had been working for months on a new Van Halen album, *Balance,* and I took a little break to get married in November 1995. We got married in Mill Valley at the amphitheater on top of Mount Tam. Beautiful, sunny day. My mom was happy. Kari's grandmother was there, her mom, all my family. My pal Emeril Lagasse, the great New Orleans chef, flew out and cooked for the wedding. We had ten pounds of white truffles imported from Italy. Ed and Al were there. Everybody posed for the picture in *People* magazine. Someone overheard someone talking to the brothers, saying, "This fucker's making too much money."

Eddie was supposed to be sober, but he wasn't and he could be trouble. He couldn't drink around Valerie, and Ray Danniels was all concerned that we keep Eddie straight. I had taken Eddie to the Bridge School concert in 1993, the all-acoustic benefit for a school for children with severe learning disabilities run by my buddy Neil Young. I did a couple of the shows by myself and I was terrified. I don't lack confidence one bit, except for when I'm by myself with an acoustic guitar, and then I'm a wreck. Neil Young is a fearless musician. He starts stomping his foot, slapping his guitar and singing at the top of his lungs. He doesn't have any inhibitions. James Taylor was sweet backstage. "Sammy, what do you mean you're nervous?" he said. "We all want to be like you."

"What do you mean," I said. "Scared?"

The year I brought Eddie, the headline act was Simon and Garfunkel. Eddie and I were both very nervous, but we did well. Three songs on piano and Eddie played a solo on this tiny amp with a kind of acoustic setup and he was great. We didn't go over as well as you might have expected, but it wasn't our crowd.

We went back to our trailer and we did some blow. Paul Simon was in the trailer next to ours and I started talking with him, while Eddie was getting more trashed. He finally wandered out to see what was happening. Paul Simon invited him to play on a song. "Do you know 'Sound of Silence'?" he said.

"No, I never heard of it," said Eddie.

Simon took him in the trailer and tried to show him the song. He was supposed to take the stage in about twenty minutes. Eddie couldn't get it. I guess he was too wasted. "Wait," he said. "What key again?"

He tried his finger-tapping to the song. Eddie's a great musician, but very methodical. He doesn't simply jam those things. He finds the melody and plays that. He was looking for the melody while Paul was singing and playing him through the song. And he couldn't get it. "Never mind, Eddie," Simon said.

"No, no, no," said Eddie, leaning over his guitar again. Simon finally dashed off to the stage and did call out Eddie, who went out there and butchered the song.

We were making the *Balance* record, but it was over for Van Halen. If it wasn't for the producer Bruce Fairbairn, we never would have finished that record. He had to throw Eddie out what seemed like every night. Eddie would come in seeming drunk and fucked up. You'd go into the bathroom in the studio and there'd be a hole in the wall. Reach down and there was a bag of cocaine. A bottle of vodka was underneath the sink. Chewing gum and cigars were everywhere. "Al," I'd say, "your brother's fucked up. What is this bullshit, everybody saying he's clean and sober? The guy's ripped out of his brain."

"You're crazy," Al would say. "That's just the way Ed acts."

I'd get in Eddie's face. "Ed, get the fuck out of here. You're fucked up. I don't want you in here while I'm working. I'm doing my vocals. Get the fuck out of here."

"I haven't had a drink for five months, you motherfucker," he said. He'd break down and cry, bust up things.

It got ugly. Fairbairn and I were staying at the Bel-Air Hotel, and, the times when Eddie became thoroughly disruptive, he would call the session and the two of us would drive back to the hotel together, sit in the bar, eat a bite, and drink a couple of cock-tails. Eddie was really on edge, because, number one, he needed a hip replacement and was taking painkillers all the time. And number two, it seemed like he was drinking and hiding it from ev-erybody. When I started to do my vocals, Eddie, for the first time ever, started making suggestions about how I should sing. That got out of hand so quickly that Fairbairn took me to Vancouver to finish my vocals by myself.

I knew they were trying to get rid of me. Eddie was trying to make me quit. He would find something wrong with every lyric

I'd write. He'd never said a word about a lyric before. Suddenly he didn't like anything.

"That's wimpy," he said. "Make it 'Don't tell me what love can do.'" I had this strong, positive thought—"I want to show you what love can do"—but Eddie wanted to switch it around. *I want black, no, I want white.* Okay, I'll go with white. *No, I want black.* Okay, I wanted black to begin with. *You know what? I want white.* It would drive me crazy. The brothers were dead-against me.

I wrote that song, "Don't Tell Me What Love Can Do," about Kurt Cobain. I wanted it to say, "I want to show you what love can do." Ed and Al fought me on that. They wanted more of a grungy, bad-attitude song. "Don't tell me what love can do." That's not what I had in mind. I was talking about somebody who could have saved Kurt Cobain's life.

I do believe that. You can save people. Drugs kill people. People think drugs are what made Jimi Hendrix great. No, drugs are what killed Jimi Hendrix. Kurt Cobain could have been saved. The people around him let him go, for some reason. They had to have seen that coming. So I wrote that song about it to say you have control over your destiny. It's your life. You can do what you want. But then I wanted the chorus to say, "But I want to show you what love can do." I wanted to make it a love song. Not about me and Kurt Cobain, but about what people could have done for him, people that he knew and loved.

Kari was pregnant, and they hated me being so happy. I kept telling Kari I had to get out of the band, but I didn't want to quit. I saw what they did to the other guy. They will lie. They will crucify me. They will kill me with the fans. The fans went against Roth. He died a quick death as a solo artist. Maybe not instantly—he had a brief moment when he first went solo—but it didn't take long. I didn't want that to happen to me.

It didn't help that my ex-brother-in-law, Bucky, died about three weeks before the Balance Tour started in March 1995. It was a heartbreaker. Bucky and Joelle had divorced, and for a while he and his son, Ben, had been living together on a houseboat in Sausalito with Bucky's new girlfriend, Penny, a great gal who used to be Jeff Beck's old lady. Bucky lived for his kid, and when Ben died in a car crash—a bunch of kids riding in the back of a pickup truck on the way to Stinson Beach and Ben was the only one killed—the bottom fell out of Bucky's life. My lawyer sued the city over the accident and came up with a fair-size settlement for Bucky. When Bucky died from an OD, they found the check crumpled in his fist.

By the time we went on tour that March, the Van Halen brothers were both in terrible shape. Al came out of the tour in a neck brace, because of a ruptured vertebra. Al collapsed in the hotel lobby the day of our dress rehearsal in Pensacola, Florida. His hands went numb and down he goes. He started seeing osteopaths every day, getting these crazy adjustments. In Paris, the doctor put on rubber gloves and stuck his hand up Al's ass to work the lower vertebrae. If that wasn't enough, he and his wife were separated, the beginning of another divorce for Al. He was under a lot of pressure.

Eddie seemed like he was on painkillers most of the time and was facing a total hip replacement due to avascular necrosis, a bone disease often associated with alcoholism. Eddie walked with a cane—his hips were shot. He would walk up to the stage, put the cane down, and walk out. Every so often, he would sit on the drum riser or a stool to play a couple of songs, because his hips were killing him so bad.

On the final leg of the Balance Tour, Ray Danniels booked the band to open for Bon Jovi at football stadiums in Europe in May and June 1995. It was a total disaster. Van Halen had no place on a bill with Bon Jovi, who was huge over there. They did three nights at Wembley Stadium in London, eighty thousand people a night, and there were about ten thousand people in the front going nuts when we played, and about sixty thousand teenyboppers in the back waiting for Bon Jovi. As soon as we finished playing, our people left and the Bon Jovi kids came to the front of the stage. It was total oil and water. Nothing against Jon Bon Jovi. He and I went to dinner many times on this tour. But it was the worst idea ever for Van Halen. We got nowhere on that tour. I could feel the end coming.

Still, Van Halen rocked. We would play a killer show, walk off-stage together, hugging and laughing at what a great show we just played, and the next day it would be back to the same shit.

We flew separately to Japan to do the last shows on that tour. We stayed in different hotels. About two o'clock in the morning, Eddie called. He had wiped out his minibar. Wasted on his ass. Clean and sober? These were almost the last shows. On the way home, we were stopping for four nights in Hawaii, but then we'd be done.

"What are you going to do when we get back?" he said.

The Ronnie Montrose story all over again.

"I don't know," I said. "Take some time off. What are you going to do?"

"I don't know yet," he said. "When I figure it out, I'll let you know. I've got some plans, but I'll let you know if it involves you or not."

"Okay," I said. "Fuck you." I hung up the phone.

We went to Hawaii to play the last shows. Kari and I decided on an impulse to buy a house. We'd been renting places every year for three months, from Thanksgiving through, like, January or February. We were on our way to the airport and I called a realtor. I told him I wanted something private, on the ocean, lots of acreage, a guest house, a pool, and total privacy. I want to be naked.

I want fruit trees. He took us to see this place on a cliff on Maui. We bought the house on the spot, and decided that, when the tour was over, we would move to Hawaii to have the baby. We were going to have this baby through natural childbirth. I wanted to deliver it. I wanted to have the baby and take a long break from the band.

As soon as the tour ended, the brothers started calling every day. We'd just gotten off tour. We just did a record and a world tour, and these crazy bastards wanted to go in and do a song for the movie *Twister*. I was not down with it. All they wanted was to get me off the island. Ray Danniels would be on the phone saying things like, "If you're not back tomorrow, we're assuming that you've quit the band."

I talked to the film director on the phone. He sent me the script. There were some key words—"drop down" was one phrase—that are used by twister-chasers. I wrote these great lyrics for a song called "Drop Down." I cut a little demo over there and the director loved it, told me I told the whole story in three minutes.

They hated it. Al and Eddie told me it was stupid to write about the movie. I told them I had been working with the director.

"He doesn't know what he's talking about," they said. "We don't like it. Get over here. If you're not here tomorrow, we're assuming you quit the band."

Kari was so pregnant that she was ready to pop, and I was flying home. I flew my mom over and I flew back. This was about the fifth time I had to fly back to the mainland. I wrote new lyrics. Bruce Fairbairn was waiting for me. Eddie wanted to call the song "Human Beings." I wrote all these belligerent lyrics—"lemmings breeding . . . There is just enough Christ in me to make me feel almost guilty . . . because we are humans, humans being."

I was ready to fly back to Hawaii the next day, but they told me they wanted me to stay and work on another song for the greatest-hits record. I told them I wasn't doing any songs for any greatest-hits record and split. I went back to my hotel room and changed my name at the front desk. I didn't want to call Kari at four in the morning and tell her. Eddie was trying to call me all night. Security knocked on my door to tell me they had Eddie Van Halen on the phone and he wants to know what room I'm in. "What do you want us to do?" the guy asked.

"Tell him to go fuck himself," I said.

That's when they called Roth.

The greatest-hits album was Ray Danniels's idea. They wanted some quick bucks. I thought if we're going into the studio, let's do a whole record, but they wanted the greatest-hits record. Then he gets another genius idea—let's get David Lee Roth back, do two new songs with him, two new songs with Sammy, and we'll be bigger than God. They did the whole thing behind my back. I was thrown out of the band for not going along with it.

I went back to Maui the next morning. Kari was way pregnant. We talked it over. In another few days, she wouldn't be able to fly anymore. We agreed going back was the best thing to do. Back in Mill Valley, we went to see the pediatrician who delivered Aaron and Andrew and he told us the baby was breech and would have to be delivered by caesarian. Forget that I was going to deliver the baby in water and all that stuff we learned at Lamaze class in the church in Hawaii. She would have to have been helicoptered to a hospital. It turned out to be a good thing that we went back.

When my first son, Aaron, was born, I wasn't even in the room, because we were on welfare. Dave Lauser and I were out in the park, eating fish and chips, and no one even told me. I finally went and checked. When Andrew was born, I was right there. He was such a surprise, because I was certain he was going to be a girl. We painted the room for a girl. We bought girl clothes. The baby shower was all girl presents. Even the doctor said it was probably a girl. I burst out laughing in joy. A child is a child and, when it's your child, it changes your life. I think it was the most joyful moment in my life. Until Kama came. When Kama came, it was even more of a joy, because I actually took her out and cut the umbilical cord.

Kari came home with our daughter from the hospital, and the next Sunday was Father's Day 1996. We were lying in bed around nine in the morning, with the baby, when the phone rang. It was Eddie Van Halen. He had been up all night.

"You've never been a team player," he said. "You never want to do things when we want to do them. You always wanted to be a solo artist. You can go back to being a solo artist. We've been working with Roth on the greatest-hits record and it's going great."

They'd been working with him behind my back while I was in the hospital with my wife having a baby. I went off.

"Fuck you, you fucking motherfuckers," I said, and hung up.

They couldn't take me being happy for one more minute. They had to get rid of me. It irritated them so bad that I was so happy. I had my little girl, my wife, and was living the happiest life on the planet. I called Ray Danniels. "Eddie just called me and said you've been working with Roth," I said.

"Oh, no," Danniels said. "He didn't make that call, did he?"

"Yes, he did, dude."

"What do you want me to do?" he asked.

"Go fuck yourself, for starters," I said. "Second of all, congratulations. You just broke up the biggest band in the world. That's going to be a big feather in your cap."

I went off on his ass. "Let me talk to him first," Danniels said. "Don't take that attitude."

"Fuck you. It's over," I said.

Eddie always said I quit, and maybe I did. His attitude was that

I always wanted to be a solo artist. They even attacked my work ethic. He and Alex told Kurt Loder of *MTV News* that I didn't want to work. I remember reading one article where Eddie said, "He was a lot older than us and I don't think he really wanted to work like we did."

I gave them their Van Halen rings trademark. They gave me my Cabo Wabo brand. I kept my royalties. I was a 30 percent partner in that band, since they'd already knocked Michael Anthony down to 10 percent.

Things that put a stick up my ass make me take action. I think sometimes I'm at my best when I have something to prove. When I joined Van Halen, I was burnt out and finished with the business. I didn't even want to be creative at that point. When I replaced their first singer and was taking shit from everybody, putting myself on the spot, it lit a fire—*I'll show these motherfuckers*. I became really driven in that band and we did some amazing things, even at the end. Even our last record, *Balance,* was a great record. I'm an adrenaline and inspiration junkie. If something inspires me, I will get up for it. With inspiration, I can do anything. When I was kicked out of Van Halen, I was determined to show those motherfuckers that they made the biggest mistake of their lives.

☆12☆

MAS TEQUILA

I was out of Van Halen. One side of me was angry, but the other was nothing but happy.

Kari and I got on a plane with our brand-new baby to go back to Maui. We had a little dog called Winchell that we snuck on board. You can't take a dog to Hawaii. They quarantine the suckers for six months. We were planning to stay a good spell—I certainly didn't have any big plans—so we pumped the little pooch full of doggie tranquilizers and stuffed him in a bag. Sitting across the aisle in the first-class cabin was Mickey Hart from the Grateful Dead. I knew who he was, but we'd never really met before. He and his wife, Caryl, were headed over to the islands for some downtime, which, it turns out, is something Mickey Hart knows nothing about.

Bill Cosby was also on the flight, about four rows behind us. It was an early-morning flight and we were snoozing as the plane was getting ready to land. Out of the corner of my ear, I hear that famous Bill Cosby voice speak up: "Oh, what a cute little dog."

I turned around and looked. There was Winchell, staggering

down the aisle, wobbling, tripping, falling like a drunk. He was a rat terrier and he dug his way out of his bag, unzipped the damn thing, and got loose. We were busted.

Mickey's wife turned out to be a lawyer and she swung into action, schmoozing the stewardess. We had to stay behind on the plane. Bill Cosby walked by me, looking at his bag and going, "Woof . . . woof." Mickey and Caryl stayed with us. We were there for a couple of hours. This was a serious deal. We were facing a possible $25,000 fine and even jail, but after a few $100 bills were passed around, a dog carrier was brought to the plane and Winchell went into the baggage compartment to fly home with the stewardess, who handed him off to some of our friends who were waiting for the dog in San Francisco.

When I got on that airplane to go to Hawaii, I had accepted that Van Halen was done. I was going to Hawaii because my place there is my sanctuary. If I went to Cabo, the press would have been all over me. What happened? What happened? I just wanted to lie back and figure out what to do. I was going to Hawaii to get my head together and decide what I really wanted. Did I really want to keep doing this? Financially, I certainly didn't have to work. I'd been doing that tour/album, tour/album grind since Montrose. I was thinking I was going to lie back, do nothing until something came to me. I wasn't looking to put a band together. I was going to hide out.

But Mickey Hart wouldn't let me.

Mickey came over to my house in Maui every day. I told him I was through with the music business. He told me I had to get right back on the horse, that I was too talented to quit. He would come over, light up big fat joints, and get me to play guitar. He had all these cassettes of African music and would be constantly snapping tapes in the deck and telling me, "Listen to this." He totally put me right back on the horse, that knucklehead.

Mickey's the most energetic guy in the world. He has never taken off five minutes in his life. He reads six newspapers a day, writes a couple chapters in a book, knocks off a couple of songs, and goes to rehearsal.

"What do you mean you're going to take some time off?" he wanted to know. For him, it wasn't even about "You've got to show those guys." It was simpler than that: "You're a musician and a singer, so that's what you do."

He has this catalog of beats and world music that he's collected over the years and carries around with him. He had Egyptian, African, South American, all these different styles of music. He kept playing all this music I'd never heard before. It was very inspiring. I picked up a guitar and started jamming and, in no time, we had written about four or five ideas. He was coming over every day, rolling up a fat one. They've got the good stuff over there, too. I didn't smoke as much as he did, but he'd get pretty high and would get me worked up.

I turned around and came back to California. I went up to Mickey's house, and the two of us would crank up this African music as loud as it would go. I played guitar and he sat down at the drums and we jammed for about three days. "Marching to Mars" was the only song that stuck, but I got interested in doing these other kinds of grooves. I was drawn back into making a record. I asked Mickey to coproduce it with me. I started thinking about putting a band together. I'd just gotten out of the damn frying pan, and Mickey dragged me right back into the fire. If he hadn't been on that airplane, I probably would have stayed in Hawaii for months.

We went into the studio and Mickey went totally crazy. He never stopped. He piled up overdub after overdub until he needed to bring in another recorder. For one track we did, "Marching to Mars," he brought in four twenty-four-track machines and used all ninety-six tracks. Hart was on the phone at four o'clock in the morning, trying to find another twenty-four-track, when engineer Mike Clink from the Guns N' Roses sessions finally said, "That's enough."

This one song took more time and was turning out more expensive than the whole rest of the album. I made the mistake of telling Mickey to stop.

"You've wasted enough time and money on this one track," I said. He got so insulted he went out and sat in his car, rolled up the windows, and lit a joint. Nobody could find him. I finally went out to the car and there he was, sulking. I apologized.

"It's four o'clock in the morning," I told him. "We're all worn out." We finished the track, the last cut on the album, which had been finished for almost three months except for this one final track. I love the guy. He may be the most high-energy, hardest-working, most enthusiastic person I've ever met.

For the album *Marching to Mars,* the band included Denny Carmassi from Montrose on drums, Bootsy Collins on bass for a couple of tracks, and John Pierce from Huey Lewis and the News on the rest. Jesse Harms played keyboards, and the engineer Mike Clink also produced a bit. I went into the studio and made the best record I could possibly make, an artsy record, a sharp left turn from Van Halen. It was one of the best solo records I've ever made. Every song is great.

I paid for the record myself and didn't want record companies involved until I was done. For the release of *Marching to Mars,* I signed a deal with a new label run by Sid Sheinberg, former head of MCA. He had retired and started this movie company called the Bubble Factory, and a new record company called the Track Factory. They gave me a large advance and big points. I was the only act on the label and they would do whatever I wanted. It was like a dream come true. We went to Hong Kong and did a press event. We went to Japan and played acoustic at a couple of in-stores.

The first week the record came out, it sold 44,000 copies—not chicken feed, but not millions. The next week, the company folded. They'd put out a big-budget movie starring Bette Midler, which bombed, my record, and that was it. They went out of business.

MCA took over, but the momentum was lost. The record was done. In the end, it sold pretty well, but it was a disappointment to me, because I came out of a band that was selling 5 to 7 million records. *Marching to Mars* sold about 400,000. I took a long fall. But it was a successful record in its own way.

The Track Factory hadn't been the only option. There had been another guy, who wanted to sign me to Hollywood Records, the Disney label. He slept on my floor for four days, trying to get me to sign a record deal. He wrote up a deal on a cocktail napkin. It was big money, way more than the other guys, but I backed out at the last minute. He was too crazy. On the cover, he was going to have a van with HAGAR painted on the side. He was going to tour the van to every record store in the country, and give it away in a contest at the end.

I decided to put a band together. I wanted somebody the opposite of Eddie Van Halen. Every guy I auditioned would try to do Eddie's five-finger tapping thing. Anytime somebody did that, they were done instantly. I wanted a black guy who played more like Hendrix or Stevie Ray Vaughan. Somebody told me that Vic Johnson of the Bus Boys was a big Montrose freak. I brought him up from Los Angeles for an audition. I asked him, did he know "Three Lock Box." "Hell, yeah," he said and off he went. I hired him on the spot. I had David Lauser back on drums, and Jesse Harms played keyboards. Jesse was real important to my music during this period. He supported my songwriting, writing bridges and choruses, and he was a soulful singer, although we eventually had a head-butt and I fired him.

I wanted a girl bass player and they are hard to find. White

Zombie had one. David Lauser found Mona Gnader. She was living way up in the sticks near Willits, California. She pulled up to my house on a Harley, with her bass strapped on the sissy bar. And she could play her ass off. One thing I loved about Mona is she's like Michael Anthony's twin. She's left-handed but she plays right-handed. She's a little fireplug, about the same size. They both have this high voice. They're like sister and brother from another mother.

The second Mona started playing in my band, I became a better singer. Most bass players play really hard, like Michael Anthony, banging it until he's knocking it out of tune. Singers get their note from the bass, whether they know it or not. You may think you're listening to the piano or the guitar, but the second that bass starts to play, you're singing to the bass. Mona has tiny fingers and she plays like Paul McCartney, very soft. She cranks up the amplifier, but she hits the strings lightly. Suddenly I sing dead on-key. Kari had to spiff up Mona a bit. Mona never owned any clothes but a pair of shorts, a pair of jeans, motorcycle boots, and a T-shirt, and she never wore lipstick or makeup in her life.

I decided I wanted to dress like Janis Joplin, so I went to Haight Street before that tour, and bought crushed-velvet stretch pants. I was going to go hippie in this band with a biker chick and a black guy. I didn't want a heavy metal, glamorous rock band. It took me a while to figure out exactly who we were, but I knew I had this great, quirky little band that I named the Waboritas—and later shortened to the Wabos—and we rehearsed every day.

We did 142 shows that year. I went to promoters named Louis Messina and Irv Zuckerman, out of St. Louis, the two guys most responsible for breaking me way back in the beginning, and arranged for them to coproduce the entire tour. I played three-thousand-seat theaters and did every city in the country, 142 shows that year and another 138 the next year. We went door-to-door. Everywhere we went, I was saying, "I am back. I am back." It was the hardest I ever worked, twice as hard as Van Halen. I kept meaning to slow down, but instead I keep stepping it up. I don't know what's wrong with me.

I tried a bus for about the first two weeks of that tour. We were playing almost every night. I'd get back on the bus and I couldn't sleep. I chartered an eight-seater turbo prop Beech 200. It was expensive for how much we were making in the theaters. I was carrying a pretty big production. I hired Jonathan Smeeton, who did all those great Peter Gabriel shows, and he knew how to take one truck's worth of gear and make it look huge. He was also a great lighting designer, but all that was expensive. I didn't really care about the money I was making on tour. I was just trying to get back in the game.

Kari loved our band and everybody loved little Kama. Vic Johnson would sit with Kama on his lap on the airplane. That's how we rolled. We all jumped on this airplane, every seat taken, tour manager sitting on the toilet in the back, and we flew all over the damn country. We were trying to write songs for a second album while we were on the road. I wanted to do it the old-fashioned way. When we weren't touring, we went down to Cabo. That's where we started to find out who we were and invent the party.

I DECIDED I wanted to make my own tequila for the cantina. I'd first tasted real tequila when I was shopping for furniture for the cantina in Guadalajara. The 100 percent agave brands were not available in the States at the time, like they are now. I'd always loved the ritual of tequila—the salt, the hit, the lime. That's fun when you're partying with friends. But you don't have to do that with good tequila. The salt is important for the first taste, to clear your palate, like having a salad before a steak. It just sets it right up. When I tasted real tequila, I flipped out.

Just finding agave growers to make it for me was difficult. Most of them sold their crops to the big manufacturers, and, if they kept any to make their own, they made small batches, like twenty cases, for their friends and families to drink. I finally found a farmer who would do it and deliver the tequila in brand-new five-gallon gas cans and plastic bottles. We transferred the tequila to oak barrels, real tequila-aging barrels that we bought, and served it right out of the barrel.

When I'd just started making the tequila, Kari and I were still going over to Maui every chance we could get, even though we were working so hard. I got reacquainted with Shep Gordon, Alice Cooper's manager, who lived in Maui and owned one of the island's great restaurants. I showed him the tequila and he liked it. He called Willie Nelson, who also has a place on the island, and Willie came over to Shep's to taste the tequila. "That's damn good tequila," he said.

I had some porcelain bottles made and we started bottling the stuff. Shep Gordon found a distributor on Hawaii and we shipped a hundred cases as a test. The corks didn't fit, the bottles cracked. Half the cases arrived upside-down. It was a mess. We started making bottles out of hand-blown glass and shipping over more cases, until we finally got it right. But our manufacturer landed in trouble with the Mexican government, who confiscated some of their property for back taxes, and they were demanding a million dollars to go ahead. We started looking for another grower.

That's when we found the Rivera family, three generations of family, the grandfather, the father, and the son, all working together in the fields. They didn't even have factories. They had mules pulling carts in the field. These guys would dig a hole in the ground, start a fire, and cook the agave right there. Their tequilas were really trippy, much smokier, but very inconsistent. Every batch was different. Every now and then, they would hit on something.

In 1999, Shep Gordon made a deal with Wilson Daniels, a high-end wine dealer. I knew who they were, since I'd been collecting fine wine since Capitol Records president Bhaskar Menon gave me a case of 1966 Pichon Lalande Bordeaux for Christmas. These guys dealt with the Échezeaux, La Tâche, Romanée-Conti, Richebourg, wines so fine and so limited in production, people are happy if they can buy a couple of bottles, never mind a couple of cases. They were interested in getting into the spirits business and ordered six thousand cases of Cabo Wabo. The Riveras had to step up to deliver. They were used to making twenty, maybe fifty cases a year, but they managed.

About this time, I ran into Narada Michael Walden, the Marin County record producer who'd made those big Whitney Houston hit records, "How Will I Know" and all that. He said he wanted to produce me and I asked him, if I let him produce me, what would he do. He told me to go out and find my favorite rock track, loop it, and write a new song. The rappers were all doing that—Tone Loc's "Funky Cold Medina" had actually used "Rock Candy." The first thing that popped into my head was "Rock and Roll Part Two" by Gary Glitter. I'm thinking, "Great fucking idea."

I asked Jesse Harms to loop it, and I wrote *"Mas* Tequila." We went into my little basement studio and Lauser played drums. Everybody at MCA got all excited. The Wabos and I made our second album, *Red Voodoo,* downstairs, crammed in, totally digging the small-time, basement studio vibe. I didn't care if the drums sounded like crap and there was leakage. If it was a good take, that was the magic I wanted. It was the opposite of *Marching to Mars.*

Shep came down to Cabo. We went to the factory. He came to the cantina. He saw the band. I had this 100 percent agave tequila that was freaking out everybody who tasted it. He saw me onstage in a bathing suit, Mona wearing shorts and flip-flops. "Roll it all together," he said.

It made sense. Take the lifestyle and bring it to the stage. It was who we were. We were getting ready to bring out the tequila. It all snapped together.

I had heard of Jimmy Buffett, but didn't really know what he was about. Kari drew the connection immediately and she took me to see a Jimmy Buffet concert at Shoreline Amphitheatre in Mountain View. I asked Jonathan Smeeton, who'd been so great on the last tour, to design a set that looked like the Cabo Wabo. He went down for a week, took pictures, made drawings, and came back with a stage. He's got the audience onstage. He's got the palm trees, the *palapa* roof, everything.

"*Mas* Tequila" comes out and is a huge hit—most adds the first week, fastest rock radio track to the top of the charts, stayed there for weeks, one of the big hits of the year in 1999. (After the song came out, I ended up with only one-third of the songwriting credit, even though I took out the loop and reversed the chord change. MCA's lawyers split the take with Gary Glitter and his songwriting partner.)

Shep Gordon talked the Hard Rock Cafe into hosting a promotional tour. He got MCA to pay. We did fourteen cities, free concerts, tied in with radio stations, the works, and we launched the tequila. We sold thirty-seven thousand cases the first year, instantly the second bestselling premium brand in the country. Something like Tanqueray gin only sells fifty thousand cases.

Looking back now, I can see I wanted to be a small-time band again, get far away from that gigantic Van Halen scale. I wanted to go back and be a club band, roll that whole Cabo Wabo vibe into everything. We loved playing down there. We'd go down there on our time off and have a blast. We'd play for free at the cantina. The place was always packed. Everybody was drunk. Nobody cared what we played. I was done with that big-time Van Halen thing.

When we went out on that tour, I opened the show by walking out in front of a closed curtain wearing shorts, shades, tank top, and flip-flops, house lights up. I'd introduce the Wabos and then I would have a waitress in a bikini bring me the fixings and make myself a cocktail. I'd finish with the tequila. "Here's the way you do it," I would say. "You put a little Cabo in there." As soon as I said "cheers," the band would break into an a cappella version of "Cabo Wabo." It was something a little different for my crowd.

That was the invention of the Wabos. We became exactly who we are. This is the way I live. This is what I eat. This is what I drink. This is how I act. This is the way I play. These are the kind of songs I sing. We found ourselves and that's when the whole birthday bash thing took off.

The Cabo Wabo became a place where anybody could come down and play. I never charged for my birthday bash. It was special to me. I brought my whole family, my brother and sisters and their families, everybody. People started showing up—Slash, Alice Cooper, Rob Zombie, Mickey Hart, Bob Weir, Stephen Stills, drummer Matt Sorum and bassist Duff McKagan of Guns N' Roses, Jerry Cantrell from Alice in Chains, Billy Duffy from the Cult, and, of course, Michael Anthony. Chad Smith, the Red Hot Chili Peppers drummer, started coming.

Toby Keith flies in every year for my birthday. Kenny Chesney came down one year with his whole band and played for three hours and forty minutes. He holds the record at the Cabo Wabo for how long he played. He wore my ass out playing "Eagles Fly," "Fall in Love Again," some of the Van Halen songs, his favorite stuff. He still claims the only reason he came off the stage was because he had to take a pee. He was drinking a lot of beer up there.

John Entwistle of the Who had a timeshare down there. His birthday was October 9, a day after my brother's. He came down every year for my birthday party—my annual birthday celebration usually lasted two weeks or more. Entwistle loved to party. We probably played together there five years in a row. The last year before he died, he came over to my house. He was so deaf. He spoke really low because his hearing aids were turned up so loud. He took them out to change the batteries and the damn things were screeching louder than the waves crashing outside my deck. I couldn't believe he didn't hear it. He put them back in like nothing happened. What a sweet man. He was pretty high most of the time. John always had a drink and a cigarette in his hand. He didn't walk around with his hands free. He wore his snakeskin boots, tight jeans and giant belt buckles with spiders on them, flashy shirts and big old shades.

I'd try to get him to sing "Boris the Spider" but he'd go, "Oh, man, I can't sing." We'd jam on Who tunes—"My Generation," "Won't Be Fooled Again," "Summertime Blues." I always played guitar when John was there. I loved playing the Who songs with John. He could really play. I never saw a set of fingers on anyone like his. He would take Mona's amp or Mikey's amp and blow them up. Every time. I've got good pictures on the cantina wall of John.

One year, Stephen Stills came down. There were a bunch of people already there—Matt Sorum, Michael Anthony, Jerry Cantrell, and a couple of the guys from Metallica, drummer Lars Ulrich and guitarist Kirk Hammet, along with my whole band.

Stephen's tour manager called ahead. I told him we would be excited to see Stephen and was there anything he likes that I could get him.

"Stephen likes coke," he said.

Stills showed up around midnight. We had already played a set—Lars Ulrich, Jerry Cantrell, and a bunch of us. He walks in wearing a tweed wool jacket. It's 110 outside. He's got long pants, boots, sweating like a maniac, dragging his overweight ass up the stairs. I'm a big fan, but this guy is fucked up. I take him in the bathroom and give him a gram of coke that I had somebody get. He opens it up, closes it back, throws it on the ground, reaches in his pocket, and pulls out a Bayer aspirin bottle full of coke. "I've got my own," he said.

He tapped out a bottle-cap for each nostril, *pow . . . pow.* I did a little. It was powerful. A guy who tried some later told me it was so strong, you touched it and your face went numb. Everybody dug in.

We went out and Stephen started playing "Crossroads." Matt Sorum played drums, Jerry Cantrell and I were playing guitar, and Michael Anthony was on bass. After a bit, Lars slid in behind the drum set and Stephen struck up "For What It's Worth." Lars didn't know the song, so he just started beating on things. Stephen stopped the song. "Where's that other drummer?" he said. "Get that other drummer down here." Lars practically crawled offstage. But Stephen was cool. He didn't care. He wanted the other drummer.

Then those guys got lost for three days. They disappeared. They went out that night, they went someplace and didn't come back. I can't hang like that. When they came back, I heard what happened. They all said Stephen took them down, all the young bucks, and showed them. "He put us all to shame," Lars said. "We saw the sun come up three times."

I tried to get with Stephen another night. I went over to this penthouse place where he was staying and took a couple of acoustic guitars. He is really a great acoustic guitar player and I wanted to learn something from him, some of his tunings, maybe cowrite a song. We got so high, by the time we picked up the guitars, it was useless. I tried to show him a song idea, and he couldn't care less. Then he would try to show me something, and I'd be like, "Okay, well, maybe, no, next." There was no connection. I love Stephen, but he's a hard guy to communicate with.

"What's with Steve?" I asked his tour manager. "He shines me on. You say something to him, he'll turn around and walk away."

"He can't hear," he said. "He probably doesn't even know you said anything."

I climbed aboard the airplane to go home—I was flying commercial—and looked across the tarmac. Here comes Stephen, limping his way to the plane, dragging his leg like the mummy. He's still wearing the tweed sport coat—he probably hasn't changed his clothes the whole time. He plopped down in the seat across the aisle, one row ahead in the first-class cabin. He didn't even acknowledge me.

Finally, he recognized me and said hello, but he was shut down, not talking. His leg was obviously hurting. Then it occurred to me—all the seafood, the dehydration, the booze, the blow—this cat has gout. I've had it. I know what it's like. He'd been eating shrimp and lobster, rock clams. I bring all that stuff into the dressing room. We have these wonderful seafood feasts. He was drinking tequila like a fish and wearing that jacket. He sweated his ass off, probably didn't drink any water. He said he was in such agony on the plane, his leg was killing him. Gout, definitely. I left him alone. When we got to customs, I ditched him completely. I didn't see myself going through customs with him. No telling what he had on him.

⁂ 13 ⁂

ENTER IRVING

My pal Johnny Barbis called me from lunch at a restaurant. "Sam, have you ever met Irving?" he said. "Let me put him on the phone."

Irving Azoff was the notorious manager of the Eagles. He was one of the most powerful figures in the music business. Barbis handed Irving the phone and he wasted no time saying the right things. "You should be making a lot more money than you're making," he said.

He seemed like a really nice guy. He told me to give him a call if I ever needed a hand, and, shortly after that, I asked him to look into a record deal I was about to sign. He came back with everything buttoned up, some nice little perks included, and when I asked him what I owed him, he told me not to worry about it. Smart guy.

The next time I was in Los Angeles, I went to his office for a meeting. I was very impressed with Irving. We were talking about my tequila business, and he said, "I know somebody who might be able to help. Let's get him on the phone."

He picks up the phone, *click, click,* he's on with someone. "Hey,

Joe, I've got Sammy Hagar sitting here. He's trying to get his tequila in Costco. Think you can help us out?"

Irving knows everybody. He's smart and he knows how to make things happen. He took me under his wing and did me right. I started making a little more money. Things start happening a little better for me. If I have a problem, I call him, boom, the problem goes away. He's got power and smarts. Next thing I know, he's making deals for me, managing me, taking his percentage. I never signed a contract. Never even shook hands. But he was very fair. He didn't charge for expenses. He would send out a guy named Tom Consolo from his management team. Consolo would fly in and out, get his own room, his own transportation, pay his own way, and Irving charged me 15 percent of the gross after production expenses. A lot of managers take more than that. I thought Irving was great.

At this point in my career, I felt comfortable doing anything. I didn't care about my so-called image.

I was open to all sorts of crazy ideas. Irving and I were sitting around his office, scheming about what to do for a tour in summer 2002. We were talking about special guests and opening acts and somebody asked if I ever thought about going out with Roth, just to piss off Van Halen and get the fans worked up.

"What a great idea," I said, "but he's never going to go for it."

"Let's see," said Irving, picking up the phone. He called up somebody, and, what do you know, Roth wants to have a meeting.

I had never met Roth, only spoken to him on the phone once, so I was surprised when this tall, statuesque rock god walked into Irving's office in full drag—big hat; shades; tight, shiny black outfit with pants that hung down over his boots. I didn't know he was so tall, but when he sat down and crossed his legs, I saw that he was wearing five-inch platform heels. I was dressed in a

T-shirt, shorts, and flip-flops. He sat stiff on the chair, trying to stay taller than everybody else. I went to the bathroom.

"There has been a lot said between us," Roth said. "Let's forget it and take it from here."

It would be the last sign of cooperation from him.

Right at the start, he rejected my suggestion that we sing a few songs together and make it a friendly thing. He envisioned something more along the lines of a WWF SmackDown. We agreed to do the tour, decided we would trade off headlining—Roth one night, me the next night—with the first date to be determined by the flip of a coin, but every day after that brought another new demand from his camp.

I knew about his business. A few years before, a friend booked him for a show in Tahoe for $10,000 and kept calling me during his show to let me hear how far off-key he was and how badly he was singing. Still Roth insisted on being paid as much as I was, even though we both knew what he'd been drawing at his solo shows. He couldn't match my box office. He wanted ten times what he was making on his own. Irving convinced me to go along.

"You're going to blow him off," Irving said. "He's got this bullshit cover band. The whole world's going to finally see that you are the better of the two. Let's go out and prove it, Sammy. Come on, don't get greedy now. It's about your future. All the promoters are going to say, 'Sammy is better than him and he's the one that's doing the business and we're going to pay him more.' You're going to double your money next time you go out by going out with Dave."

We held a press conference and flipped the coin, but Roth kept renegotiating. He insisted on closing the shows in Los Angeles and New York. Irving finessed that by booking the show into a smaller venue in Los Angeles for two nights, so we could each headline one night, but the booking at Jones Beach in New York fell apart. Roth wouldn't give in and I refused to let him win. We skipped New York on the tour.

It was like that the whole time. Roth wasn't going over all that great. He lost his voice and could no longer sing very well anyway. His lame band played all old Van Halen songs, and the Wabos and I were ripping it up with new songs and my solo material, saving four or five Van Halen songs to play with Michael Anthony. The fans loved that. Roth never once asked Mike to play with him. My T-shirts were outselling his by more than four to one.

Ted Nugent and Kid Rock stood onstage to introduce me in Detroit. After the show, Kid Rock dragged me into Roth's dressing room and asked why we weren't singing some songs together. He said we were ripping off the fans by not doing it. Roth agreed to do it and we shook hands. When I sent the tour manager to see him the next night about what songs we were going to do, he came back and reported that Roth was in a bad mood and wouldn't come off the bus. When I went up to Roth later, he said he stayed up all night with Kid Rock and couldn't sing.

"My throat," he said. "Tomorrow."

He repeated that routine again and again. It got to be a running joke. I'd beat on his dressing-room door and yell, "Hey, Dave, what are we going to do tonight?"

He would go to any lengths to grab the spotlight. The *Los Angeles Times* sent a reporter to write a feature about the tour to St. Louis, my number-one market. Roth refused to give the guy an interview until right before I went onstage. The reporter decided to watch my show instead.

He pulled bullshit like that all along the tour. We rolled into Fresno and Roth called to say his bus broke down. I either had to go ahead and open the show or wait until after midnight, past curfew, to headline. We went out and opened the show and he pulled up in time to take the stage after us. What a jerk. Instead of showing any gratitude for the big business we were doing, pulling him out of nowhere and putting him back on arena stages, he was impossible. I finally shot my mouth off to a reporter for the Page Six column of the *New York Post*.

"He's a fucking bald-headed asshole," I said, "a swaggering, middle-aged prima donna who was out there pretending to be something he no longer was. He's a nostalgia act who has to wear a wig and he even spray-paints that."

We were at the Verizon Wireless Amphitheatre in Charlotte, North Carolina, when Roth saw the piece. David Lauser was walking through the dressing room, wrapped in a towel, on his way to the showers. "Hello, ladies," he said.

"Fuck you," Roth screamed. "You calling me a faggot? You fucking fag."

I stepped out of my dressing room when I heard the yelling. "Dave, you need to lighten up," I said.

"Fuck you," he said.

A couple of the road crew jumped between us, but Roth traveled with five large bodyguards and they waded in, knocking the roadies to the floor. We called the cops. After that, a plywood barrier was erected to divide the dressing room at every gig. He wasn't allowed on the premises until I was finished and I couldn't show up until he was done.

I had originally hoped that Roth and I going out together would jar those lame-brains from Amsterdam into joining up for a stadium tour by Sam and Dave and Van Halen. It would have been the biggest tour in the world. But that was never going to happen. The Sam and Dave tour was a huge financial success, but a personal disaster.

RED

* * *

DESPITE THE FRUSTRATION of the Roth tour, I kept trying new opportunities. I jumped at the chance when I was invited to play with the Dead on Valentine's Day 2003 at the Warfield Theater in San Francisco, the first time those guys got back together and used that name since Jerry died. Kari and I went to dinner before, so we arrived a little overdressed, but just seeing the marquee reading THE DEAD was heavy to me. I went over to Bob Weir's house a few days before and he told me to pick a song to jam at the show. I chose "Loose Lucy."

We talked it over on the break backstage. Phil Lesh was really cool. When I asked about the arrangement, Phil told me to just feel it. I asked how many bars before I came in singing. "Come in when you want," he said. "When you come in, that'll be the verse."

I'd sat in with these guys years before, when they were the Other Ones, but I just played guitar on the last song, "Fire on the Mountain." This time they were backing me, big difference. We did a great version—Deadhead tape traders love it. I sang a verse and let it rest. I looked around to see if anybody was going to cue me. These guys were off in their own vibe. They didn't care. When I came back in, everybody fell behind me. I really felt what they do. I didn't think it was one of those goose-bump moments, but the audience accepted me. It wasn't like when they introduced me, I got a big roar of recognition, more like, "Huh, what's he doing here?" But after I started singing, I could see they were digging it. When I came off, Mountain Girl came up and gave me a big hug. I'd never met Garcia's old lady, but I knew who she was.

"Sammy, you owned that song," she said.

For the summer of 2003, I made plans to go out with Lynyrd Skynyrd, "Party of a Lifetime" they called the tour, only to have surviving Skynyrd guitarist Gary Rossington collapse at the start of the tour from heart problems. The band canceled a string of dates. I came up with the bright idea of putting Montrose back together as my special guests and going ahead with the shows on my own. I offered the three of those guys ten grand a night to split, all expenses paid, private jet, road manager, the works. Bill Church and Denny Carmassi jumped at the chance. Ronnie Montrose was less eager.

"Okay, Hagar," he said, "but you sure you got the private plane?"

It turned out pretty great. I did my whole show and came back for the encore with Montrose. We did "Rock the Nation," "Bad Motor Scooter," "Rock Candy," "Space Station Number 5," and we always got an encore. They got paid more money than they'd ever made, and there were moments when we were really fucking good. But Ronnie started ego-tripping with my band, trying to tell my guitarist, Vic, where he could put his gear onstage, stupid shit like that. It was inevitable, I suppose, but I only did it a couple of more times.

But Irving didn't leave it there. He wanted some kind of Van Halen reunion and, since I looked like the sane one in the bunch and I was the guy he managed, he started working behind the scenes to make that happen. He got Al to give me a call on New Year's Eve 2003—the dawn of what turned out to be a very big year.

I love Al, always have. Even after I left the band, he and I would sometimes call each other on our birthdays or New Year's Eve. It was our way of staying in touch even when things were bad. The brothers had accomplished very little since I left.

They'd made an album with Gary Cherone—who'd later told me that they had auditioned him while I was still in the band. Ray Danniels managed Cherone, because he managed Extreme. First, he tried to get Cherone into *Phantom of the Opera* on Broadway. Later, Danniels told him he was going to be the singer in Van

Halen. Gary's a talented guy. Good singer, good physical shape, a healthy guy, not a druggie, really a cool guy. Wrong for the band? A hundred times over.

The album, *Van Halen III*, was the only Van Halen album that didn't go platinum. According to Gary, Eddie insisted he do exactly what he was told on the record. He told him what melodies to sing and even wrote some of the lyrics. He had never done any of that before. I remember Ray Danniels telling me, "Eddie wants his band back." I heard Eddie fired Al twice during the making of *Van Halen III*. Eddie played the drums. I always told him he should do a solo record. It took years to make the record, because of the condition that Ed and Al were in.

I don't know how they managed to tour, even the short one they did. Al couldn't play very long. Eddie was hobbling. The tour for *Van Halen III* didn't do great business. They canceled a lot of dates. They did sixteen hundred people in Sacramento. I heard they played forty minutes and Eddie walked off the stage. I wasn't there, but Gary told me after we'd become friends. I brought him to play with me at a free concert at New York's Irving Plaza I did for the firefighters after the World Trade Center went down. He said Eddie also walked off the stage in the middle of the show in Boston and didn't come back for a half hour.

They fired Cherone after the tour and started trying to get back again with Roth, which didn't last. They tried about five abortions with Roth. They would decide to get together, book a tour or start working on new material, but nothing ever happened. I knew what was going on. I kept in touch with Michael Anthony.

Meanwhile, I was doing well with the tequila. The Cabo Wabo Cantina had turned into an oil well pumping money, and so it wasn't like I needed the dough from a reunion tour. But the brothers were a different matter. They told me they were almost broke.

Al had gotten a divorce and lost a lot. When he got divorced, he was largely in debt, but Al had been deep in debt when I left the band. They'd made some bad business decisions. They were kind of low on funds and they needed the money.

When Al called me on New Year's Eve 2003, I told him on the phone that Kari, the kids, and I were coming down to stay at Laguna Beach and he should come visit. He brought his new wife and their kid. They showed up around noon and stayed until midnight. We laughed, joked, and drank. Al drank coffee and I had a couple of glasses of wine. Late in the evening, Al's phone rang and it was Ed. He flipped the phone to me. Ed started drilling me.

"Why did you quit the band?" he said.

It was late at night. I figured the guy was wasted and shined it on.

⚔14⚔

SAMURAI HAIR

I had been waiting at Eddie's 5150 Studios for more than an hour when he finally showed up. I hadn't seen him in ten years. He looked like he hadn't bathed in a week. He certainly hadn't changed his clothes in at least that long. He wasn't wearing a shirt. He had a giant overcoat and army pants, tattered and ripped at the cuffs, held up with a piece of rope. I'd never seen him so skinny in my life. He was missing a number of teeth and the ones he had left were black. His boots were so worn out he had gaffer's tape wrapped around them and his big toe still stuck out.

He walked up to me, hunched over like a little old man, a cigarette in his mouth. He had a third of his tongue removed because of cancer and he spoke with a slight lisp.

"Are you all right, man?" he said.

"I'm fine," I said.

"Well, you look a little beat up," he said.

I glanced at Al, who was laughing. Kind of. The thought flashed through my mind that I should get the fuck out of there—this guy is crazier than a loon. But he gave me the give-me-a-hug move, an awkward embrace for sure. He was not only the weirdest I'd ever seen him, he was more tore up than anybody I ever saw. But I hugged him. My idea was, if we were going to get along, we would make a new record.

"Let's go play some music," I said. "Play me some stuff. What do you guys got?"

He went digging through all these tapes and played me a bunch of song ideas, just him and Al jamming, like always. Some of it was really cool. I was going, "I like that, I like this, I don't necessarily like that." I stayed down there for a few days and tried to write with them.

The earliest the day would start was noon. There were times Eddie did not come down to that studio until nine o'clock at night. He lived next door. Al would go check on him.

"He had a hard time last night," Al would say. "He was up trying to write songs."

He may have lost a chunk of his tongue to cancer, but he was still smoking cigarettes. He claimed the cancer came from putting the guitar pick in his mouth while he used his fingers to play. I told him cigarettes killed our manager, Ed Leffler, but he didn't buy that. He walked around all day drinking cheap Shiraz straight out of the bottle. That's why his teeth were all black. "Ed, why don't you get a glass for that?" I said.

He held up the bottle. "It's in a glass," he said.

He was living with a pathologist, who kept taking slices off his tongue, to check for cancer. He beat the cancer. He told me he cured himself by having pieces of his tongue liquefied and injected into his body. He also told me when he had his hip replacement, he stayed awake through the operation and helped the doctors drill the hole. What a fruitcake.

I don't know what he was doing, but he would keep going for what seemed like three or four days at a time. He used to hang out with one of our opening acts on the tour and come into their dressing room before the show. Whatever he was doing, he kept it out of view. I never saw what it was, but he was doing something. Plus drinking wine all day. He would never be in one place longer than twenty minutes.

"I'll be right back," he would say. "I gotta take a shit." Gary Cherone told me he did that once in the middle of a show.

His marriage was over. Valerie was gone. He finally invited me over to this giant, extravagant, sixteen-thousand-square-foot house that he and Valerie had built before she split. It looked like vampires lived there. There were bottles and cans all over the floor. The handle was broken off the refrigerator door. It was like a bum shack. There were spider webs everywhere. He had big blankets thrown over the windows. The mattresses were stripped off the beds and leaned against the wall for soundproofing. He was making music and trying to get the sound right. He said we were going to record a lot over there. He had dug a trench to run wires from the studio to his house. We never used it even once for the three songs we eventually did record.

He was sleeping on the floor with a blanket and a pillow. There was no food in the cupboards. I had never seen a dirtier place in my life. It was like the house out of that movie *Grey Gardens*.

This was Eddie Van Halen, one of the sweetest guys I ever met. He had turned into the weirdest fuck I'd ever seen, crude, rude, and unkempt. I should have walked, but Eddie's got a very charming, cunning side to him, where you feel like he's got a good heart. He's going to come through. He's going to clean up and we're going to get this thing done.

I thought some of the music was great, but it was all recordings. Getting him to actually play music proved more difficult. He started to play the song that became "Up for Breakfast" on the *Greatest Hits* record. The keyboard part was already digitally recorded. Al and Eddie were going to play live to show me. I had a microphone in my hand. I was ready to jam with them like we always did. He started, and stopped.

"I've got to play the keyboard part," he said.

Ed would start the song, and then go, "Wait, wait, wait. I gotta change my amp." He'd never get more than a couple bars into it. "Oh, no, no, wait a minute. This ain't right. I gotta switch guitars." He couldn't make it through the damn song. About two hours later, Al pulled out a tape and played me an already recorded version. I loved what I heard. The keyboard sequence reminded me of "Why Can't This Be Love" and another one of my favorites, "Mine All Mine."

But the sessions were a mess. Al was in complete denial. I would try and talk to Al about his brother, but he wouldn't hear it.

"You know him," Al would say, pointing at Eddie's signature painted guitar. "See all those stripes, whacked out things, all over the place? That's the way his mind works. Everything this way, that way, scattered. Can't focus. Can't concentrate."

We planned on recording an album in three months, but pretty quickly it became apparent we had three songs that were going to be all right, and there wasn't going to be any album. We brought in producer Glen Ballard, who made Alanis Morrisette's *Jagged Little Pill,* a total pro who really tried to make things happen. I had written my lyrics. Eddie had piles of cassettes. We salvaged old tapes—sessions with Cherone? Roth? I don't know—sliced and diced them into new songs and I wrote lyrics. We had all that ready within one week. It took three months for Ed to do the guitar parts to three songs and a couple of solos. The Eddie Van Halen I first met could have done that in an hour.

I was in there one day when Ed came in with a C-clamp. Glen Ballard asked me to come down and be part of all the sessions, try to keep the vibe good and help orchestrate this thing—help him, really. He didn't want to be in there alone with this madman.

Eddie had this Telecaster he wanted to play, but he could never get it to stay in tune. Every time he tried it, it played differently. They had been working on this for three or four days when Eddie took the guitar to his workshop, C-clamped it to his workbench, and ran cables out of the studio down the driveway. It didn't work, of course.

It seemed like whenever I went to the studio around five o'clock in the afternoon, Ed still hadn't shown up yet. I'd hang around, but by the time Eddie made it down to the studio at nine o'clock, I was gone. He would burn out everybody, staying up working all night. When I tried to talk to him about it, he looked at me like I was crazy. "You know I can't do anything unless I'm creative," he said.

Ed decided he wanted to play bass. He wouldn't let Mike play bass on these three songs. On one song, "It's About Time," it took him at least a week to do the bass part. Mike could have done it in an hour. When they finally laid down the rhythm guitar tracks, that was all I needed to sing. I didn't need his guitar solos. I didn't need any of the other production. As soon as they got that, I went in and knocked out my vocals on all three songs in two hours. Michael Anthony came in and we did all the backgrounds in another two hours. Half a day, we were done. Eddie was still asleep. By the time he came down, we were finished. I left. They spent the next three months doing Eddie's guitars.

We started rehearsals for the tour. The concert was the hottest attraction of the summer. Promoters laid down big bucks and snapped up eighty dates. The Van Halens wanted a new bass player. I told them I would not do a reunion without Mike. They still managed to grind him down to a small percentage of what he would have earned as a full partner—Al, Irving, and I all gave up pieces to give Mikey some—and he signed away all further rights to Van Halen after the tour. They were mad at him. They were mad at me, too. I just didn't realize yet how mad they were.

They were still pissed about Cabo Wabo, and holding on to the idea that I screwed them somehow on the deal. If anybody did that, it was Ray Danniels, their former manager. When the government put the yellow ribbon around the club, they wanted to shut it down and wouldn't spend any money to keep it open. We were still fighting about that when the band broke up. Ray Danniels gave me Cabo Wabo in exchange for my interest in the Van Halen trademark. At the time, they didn't care, but by the re-union, the tequila was everywhere and people were always coming up and talking about the cantina. It drove them nuts.

They put it in the contract that I could not wear any Cabo Wabo T-shirts onstage or mention anything about the tequila or the cantina on the mike. I went straight to a tattoo parlor and I got this giant Cabo Wabo tattoo on my shoulder. I knew we would be carrying giant video screens and that, as the lead vocalist, I was going to get plenty of close-ups on those giant video screens. I didn't need a T-shirt.

They didn't know I did this. I made a point of wearing long-sleeved shirts. But Mikey knew. He would walk into a rehearsal and slap me on the shoulder. "How ya doin'?" he said. It hurt like a mother with all that black in the tattoo. Mike kept slapping me on the shoulder and making jokes, but the Van Halens didn't know what I'd done.

Rehearsals didn't go well. Eddie was having trouble finishing songs. Something would go wrong with his equipment. It was the same routine as when we'd first started messing around in the studio months before. He would start songs, but wouldn't finish. I left at dinnertime. Al would stay up all night with him. Eddie never played the whole set at rehearsal. All he wanted to play at rehearsal was the three new songs. He wouldn't learn the old songs. Something was always wrong. I'd walk into a rehearsal and he'd be tearing apart his speakers. The three new songs were all he

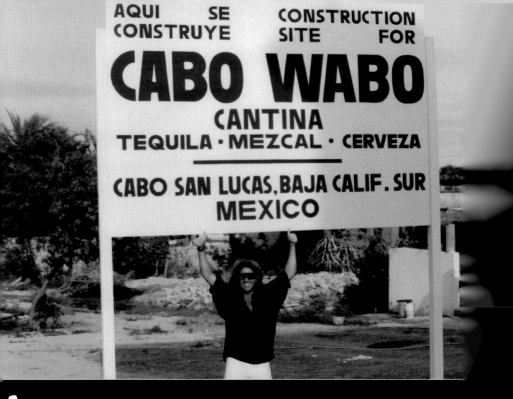

At the construction site for the Cabo Wabo Cantina in 1989.

My Merlin 3 plane that

The Cabo Wabo grand opening weekend.

For the grand opening of the cantina, Van Halen came down for the weekend and performed.

With Eddie on the Right Here, Right Now Tour in 1991.

Ed Leffler and his second wife.

With my brother, Bobby Jr., in Malibu on Grammy night, 1992.

Kari and me at our wedding dinner with the Van Halen members and their wives.

With Kari and Kama on Kama's first day home from the hospital after she was born in 1996.

With John Entwistle from the Who at my Cabo Wabo birthday bash in 2000.

At the Cabo Wabo plant in Jalisco, Mexico.

With my daughter Samantha in 2001.

At the Cabo Wabo agave fields in Jalisco.

With the Wabos in Sacramento, 2009.

The party onstage.

Lars Ulrich of Metallica at my Cabo Wabo birthday bash.

Singing with my son Aaron at the Cabo Wabo in Lake Tahoe.

F*rom left to right:* Mickey Hart, me, Bob Weir, and Mike Anthony.

Jamming with Mikey and drummer Matt Sorum at my Cabo Wabo birthday bash, 2008.

The backstreets of Cabo.

With Kari in our Cabo front yard.

At my mom's seventy-fifth birthday party in Cabo. *From left to right:* me; my brother, Bobby Jr.; my mom; and my sisters, Velma and Bobbi.

Aaron's family (including my two grandchildren) and my family together in 2009.

The Cabo tattoo.

Eddie's samurai hair during the reunion tour with Van Halen. *(Photograph courtesy of Getty Images)*

With Joe Satriani from Chickenfoot.

With Chad Smith at Cabo Wabo.

Chickenfoot at the Cabo Wabo Cantina in Lake Tahoe, 2009.

Fun with machetes in Maui, 2010. *(Photograph by ShootingStarsPhotography.com)*

At home in Maui, 2010. *(Photograph by ShootingStarsPhotography.com)*

Outside Skywalker Sound studios with my custom-made Ferrari Fiorano.

knew, kind of, and he didn't know them all that well. We could play the keyboard songs—with the keyboards on tape—and he could noodle along, "When It's Love," songs like that. We never got through the other songs—"Runaround," "Top of the World," "Finish What You Started." He could not play "Why Can't This Be Love."

He fired the monitor guy, fired the sound guy, fired the keyboard tech, fired at least five guitar techs, and that was just during rehearsals. Something is wrong when a guy blames everybody else—like the keyboard guy, who's just hitting a button that triggers the keyboard part. It was the craziest, most whacked-out stuff. I knew it was a disaster. I told Irving.

Irving is really a professional. He knows how to get things done, but Irving is not a confrontational guy. He preferred to schmooze things, but right after we started rehearsals, Irving agreed to hold an intervention with Eddie. He brought a big, beefy security guard and met Al and me at 5150. Eddie walked in, carrying his wine bottle. Irving did all the talking. He told Eddie the tour was going to be difficult, that he needed to go away for a week or two, that we could postpone some dates if we needed. We all agreed Eddie needed to clean up.

He smashed the bottle. "Fuck you," he said. "I will kill the first motherfucker that tries to take this bottle away from me. I left my family for this shit. You think I'm going to fucking do this for you guys?"

That's how sick the cat was at that moment. It was going to be a long tour.

The opening show in Greensboro, North Carolina, was phenomenal. Eddie wasn't phenomenal, but he was okay. David Fisher designed a set from an idea by Al and me that used the Van Halen rings as a way to put special seating sections in the middle of the stage. The first time I stepped out on that stage, it blew me away—the band was so powerful, the fans were so great. That carried me a long way.

But from the start of the tour, I couldn't listen to Eddie. He made some terrible mistakes and it seemed like he couldn't remember the songs. He would just hit the whammy bar and go *wheedle-wheedle-whee*. I'd listen to Mikey to find my note.

Whenever he came out with no shirt and his hair tied up samurai-style, he seemed fucked up. That was his little signal. I don't know what it was. He would come out first with his hair down, go back to change guitars, or after Al's drum solo, and come back with his hair up and shirt off. I'd look at Mike and we'd roll our eyes—here we go. Some nights he'd come out at the beginning of the show with his shirt off and hair up.

He looked like a bum in the street. His hair was matted. One time we got on a plane after a show and he spent practically the whole flight in the bathroom. When he finally came out, he had this hairbrush, the kind with the fur bristles, twisted up in his hair, hanging down. He was soaking wet, covered in water, like he tried to take a bath in the airplane sink. I made Kari look at me. I didn't want the guy in my face. He flopped down on the floor, fussing with the brush caught in his hair, and never went back to his seat, landed that way. Hospital-crazy.

When we didn't have our kids out on the road with us, Kari and I shared this big Gulfstream jet with Eddie and his girlfriend, Al and his wife, Mikey, and some management and security people. After one show, Mike and I stayed back, like we normally did, and showered. Ed didn't shower. He jumped into the limo right off the stage and went straight to the airplane. When Mike and I rolled up, laughing, joking, eating a couple of barbecue sandwiches we had ordered, Eddie was sitting there drinking his wine out of the bottle. He went off on us.

"Don't ever fucking make me wait," he said. "Without me,

you're nothing. You need me. You'll see. At the end of this tour, you guys will have nothing. You're going to have to call me if you ever want to tour again."

He was facing one direction, I was facing the other. I turned around and said, "Ed, shut the fuck up, man. Come on. We just did a gig."

"Fuck you," he said, and started bashing his bottle on the plane window. One of the security guys tried to calm him down, but he kept yelling and pounding the bottle. I turned my back on him. My guy from Irving's office was looking at me, shaking his head and zipping his lips. The stewardess and the pilot started freaking out. They were reluctant to take off with this madman on the plane. Finally Al got him to take it easy and we took off.

When we arrived at the next hotel, Eddie started asking everybody what my room number was. He didn't know the alias I used when I checked into hotels. The tour manager reached me on the room phone and told me Eddie was looking for me.

"Bring that motherfucker over here," I said. "I quit. This is done. I'm going home tomorrow. I'm not going to work with this guy ever again. He's trying to bust the window out of a fucking $40 million jet. He's got no respect for anything or anybody. Fuck him. I'm done with this tour."

I called my attorney. He wasn't very happy with the contract I had signed, once he read it. Irving and his attorney had drawn it up. Outside of dire medical emergencies, if I canceled any shows, I was liable for all lost income. He thought it could cost $5 million to leave before I finished the shows. I was trapped. Eddie apologized, but I was never flying in a plane with him again.

It was the Sam and Dave tour all over again, only it was Sam and Eddie. They kept us apart as much as they could. Irving knew better. We flew in different jets. We stayed at different hotels. We had our own limos. They had their bodyguards. Mike and I had ours. I stayed in my own dressing room on the other side of the hall. The only time I saw that guy was when we stepped out onstage. Once in a while I'd go over to his dressing room before the show and see how he was and the times I did that it was usually great. He'd start playing, I'd start singing, jamming around, like old times. Other times, he'd start telling me crazy shit, like, "I pulled my own tooth—this thing was bugging me so I got a pair of pliers and pulled it out."

I didn't think he could make it. I kept thinking each week would be the last. He was going to land in the hospital. He collapsed a couple of times. He told us one time that he had been hit by a car. He was lying down, and he was so fucked up, he couldn't get up.

"I got hit by a car," he said. "You guys don't understand."

He would go until he collapsed. Then he would pass out for a day or two in a hotel. He would wear the same clothes for a week. He would run offstage and not change, go straight up to his room. The next morning, he would be wearing the same clothes. That night onstage—same clothes. He wore those boots with the tape around them the whole tour.

His solo turned into a disaster. It used to be the highlight of every show. Now he would play nothing, just garbage. He would try to play "Eruption," one of his greatest pieces, and screw it up. He would just grab the whammy bar, hit the sustainer, and start making all this noise. The audience wasn't buying it, either. I saw his solo many nights. He would say unbelievable things to the audience. "I'm just fucking around," he would say. "I love you people. You pay my rent."

This got so bad Al threw drumsticks at him once. Another time he couldn't even stand up—he sat on the drum riser. Al had dropped a stick. He picked up the drumstick and started using that on his guitar solo. It was like a little kid banging on things.

I didn't go near him onstage. No more Jimmy Page and Robert

Plant. If he's over there, I'm over here. When he comes over here, I'm going over there. No bad vibes, just no vibes.

It seemed to me that Ed was going through the motions, like he didn't care about his playing. He didn't care about the way he looked. He just went out there and took the money. He was embarrassing. Al, Mike, and I did it from the heart. We played our asses off every night. Ed went out there and jerked off.

We went through three sound guys. He would take a board mix after shows and listen to it. The sound guy would bury his guitar because he was playing so bad. He was playing so loud on-stage anyway, he probably didn't need to have his guitar pumped through the main house speakers, but he would crucify the sound-man and fire him the next day. Al and I would argue to get the guy back, but that never worked.

Those two often shared a high school mentality. They hated every other band. It was always competitive with them. Everybody else sucked. I don't like everything, but I like music, and when I hear a musician I like, I want to embrace him, bring him back-stage, make him welcome. Eddie was usually a rude wise-guy. I brought Kenny Chesney backstage on that tour and took him to meet Eddie. Eddie shook his hand and turned around.

"I gotta take a shit," he said. He walked into the john with the guy standing right there.

"Let's get out of here," Kenny said.

That was their only meeting. It was the first night I met Kenny. We went back to my dressing room and played acoustic guitars, singing "I'll Fall in Love Again," "Eagles Fly," and all these songs of mine that he loved. We were drinking tequila and singing until three in the morning. He became one of my dear friends. But Ed? "I gotta take a shit." That usually meant he was going to go and tie up his hair.

Another time, Toby Keith came to see us in Oklahoma City, not far from his hometown. I decided to do his "I Love This Bar" during my acoustic segment and worked up this whole deal with Toby. I was going to say that since he was from around these parts, I was going to do one of his songs, even though I knew he was out of town. Then he would walk out midway through the song and sing the rest of it with me. Toby told me later that while he was waiting backstage, Eddie cornered him and tried to keep him from going out. "Why would you want to go on with him?" Toby said Ed asked. "Why didn't you come out with us?"

"You didn't invite me," Toby said.

"I'm inviting you now," Eddie said. "Why are you wearing that cowboy hat?"

"I'm a country guy," said Toby.

"No, it's because you're bald," said Eddie.

Toby walked out onstage halfway through the song and the place exploded. Eddie went crazy the rest of the night. He destroyed his dressing room after the show. His son, Wolfie, was in my dressing room, scared and crying. I went to see if I could calm him down. We left Ed behind that night in Oklahoma City with his tour manager and a couple of security guys and went to the next city without him. When they took him back to the hotel, he kicked out the limousine window.

"That boy needs help," said Toby, who drove down to the gig with his wife and teenage daughter in his truck.

Irving would come out a lot, but he wouldn't go near Ed. No one wanted to go near him, because they figured it would blow up the whole tour if Eddie quit. I'm sure the contract was written the same way for him as it was for me, if he quit or went down. If he had missed three consecutive performances, I could have walked. He never missed one. Ed never lost his work ethic. The Van Halens come from good, hardworking Dutch stock. He was there every night, in the worst shape you could imagine, but he did the show.

He was starting to let his anger toward me show. We sold these deluxe thousand-dollar packages that not only included the special seating in the stage, but you could go backstage, watch sound check, and eat at catering. I never do sound checks. I'm a singer. I save my voice for the show. But some of my fans bought these packages and showed up wearing Cabo Wabo T-shirts. Mikey told me that Eddie would pick on them. "Where'd you get that shirt?" he'd say. "What a piece of shit."

The last two shows were at a small amphitheater in Tucson. The second night, Eddie unwound completely. He knew it was the end of the tour. He knew he was done. He came up to me before the show, when I was talking to Irving, and rolled my sleeve down over my tattoo. I didn't even acknowledge him. I just rolled it back up. He rolled it back down. I rolled it back up.

"Don't be fucking with my shirt, dude," I said.

"That thing ain't gonna last," he said, showing me his Van Halen tattoo. "See that? That's better. That's going to last longer."

Like I cared. We had a crew on that tour of more than 120. I had a bunch of cases of tequila in my dressing room and I was sitting in my dressing room signing bottles for the crew. Eddie came in and saw what I was doing.

"Can I have a bottle?" he said.

I went over to my refrigerator and pulled one out. "I'll give you a bottle," I said. "These others are all signed for the crew."

He takes a couple of big slugs and sets it down. "Why can't I have one of these?" he said. I told him those bottles were for the crew and I had the exact right number. If you take one, I told him, somebody's not going to get one. He walks away, over to one of my guests in the dressing room, a booking agent Eddie knew but mistook for the son of Warner Bros. Records chairman Mo Ostin. He proceeded to give this guy a ration of shit about something that made no sense to anyone but Ed. "And your dad, he was a great man, and you and your brother are nothing."

He was raving crazy. He had already attacked Valerie's brother, who made the mistake of showing up at the concert to see his ex-brother-in-law. People were screaming and yelling in the dressing room, and he was running wild, beating up people and smashing bottles against the wall. He lost it completely.

Irving took me aside. "When this show's over," he told me, "I'm getting you in a limo and we're getting out of here." My plane was waiting to take me home.

It was the worst show we'd ever done in our lives. Eddie played so bad. My nephew was standing on the side of the stage with me, watching Eddie do his solo.

"I've never seen anything like this," he said. "What's wrong with him?"

He smashed his favorite guitar to pieces. Sprayed shrapnel into the crowd. He got on the microphone, crying. "You don't understand," he said. "You people pay my rent. I love you people."

They tell me he pulled some crazy shit on the plane home. He had his girlfriend and her two grown daughters with him. Al was there with his family. Mike and his wife stayed over in Tucson rather than fly with Ed. Some funky shit went down on that plane. My man was completely gone and out of it. I went straight to my plane after the show and home to San Francisco. I never spoke to him again after telling him to keep his hand off my shirt.

✴15✴

GOING HOME

We moved the whole family to Mexico before the 2005 school year started. Our daughters, Kama and Samantha, were each six months apart from the daughters of my Cabo Wabo partner, Marco Monroy. We were neighbors in Cabo and our kids were pals. We wanted to put the girls in school while they were still young enough to learn the language and soak up the culture. Kama was in fourth grade and her younger sister, Samantha, who was four years old, was in preschool. After the last year of Van Halen torture, I was ready for the beach.

We'd wake up in the morning to the waves crashing outside the window. The weather was fantastic. We had the Cabo Wabo and we could go down to the cantina and eat or have food delivered to the house. We had people working all over our home—maids, security, gardeners. Down there, everybody needs a job and they work hard for the money. Life can be very comfortable.

When you go on vacation for ten days, you spend the first seven days just getting relaxed. When you go on vacation to stay, you go through periods of boredom, where you break through to new levels of relaxation. You slide all the way down. When we returned after spending most of the year in Mexico, we'd changed. It wasn't just our clothes, although Kari and I noticed when we stopped at the store on the way home that the clothes we were wearing looked a little ragged and dirty back in Marin County. We'd reached a place that had a lasting effect. When we returned to the city and got busy again, we now knew where that place was, and it was easier to get back there.

During the reunion, I kept the Wabos on full salary. The only gigs we'd done all year were my birthday bash at Cabo Wabo and the annual weekend at the new Cabo Wabo Cantina that opened in Harrah's Lake Tahoe in May 2004. Ted Nugent, Toby Keith, and Bob Weir came opening weekend to play with me in the former South Shore Room, the big showcase off the casino where Elvis Presley and Frank Sinatra performed.

The Lake Tahoe cantina was part of an expansion plan I had been working on for a while. Several years earlier, Don Marrandino, who worked for the Fertitta brothers and the Station Casinos, approached me about building a Cabo Wabo in Las Vegas. The Fertittas were great. I got to know them pretty well. The Fertittas bought the Ultimate Fighting Championship, a kind of extreme boxing that mixed martial arts and no-holds-barred wrestling with boxing. They booked their first big match at the Trump Arena in Atlantic City, and I went with them. They sent a G4 private jet to pick up first me in San Francisco, and then the rest of the party in Vegas. Atlantic City was shut down in a blizzard and we landed instead in Philadelphia. They put us in big suites at the Ritz-Carlton. We went out to dinner at some fancy Italian place and they just ordered the entire menu and cases of fine wine. The next morning, we flew into Atlantic City and went to their first fights. Their father started the Station Casinos. He was the first independent casino operator outside of the Strip. He started small, but he eventually ran eleven casinos, making more money than he would on the Strip. I was thinking these guys were the smartest people in Vegas.

Marrandino came to my house to show me the plans. He wanted to do a Cabo Wabo complex—an eight-thousand-seat arena, a bowling alley, and the cantina. We were scheduled to break ground in October 2001, but after 9/11, they changed their minds. The contract ran out. Don Marrandino went to work first for the Hard Rock Cafe, and then for Steve Wynn to build the Wynn. Marrandino ended up running Harrah's at Lake Tahoe and immediately started plans to open a Cabo Wabo there. He's got friends in the music industry, and he hired them to come and play. He knew how to make a place cool and hip. Most of these old casino guys, they don't know what the hell to do. "Where's Sammy and Frank? The good guys are dead. We've got nobody to play here." Marrandino knew there was a whole new breed of people out there.

Originally, I only wanted there to be one Cabo Wabo. I hated Planet Hollywood, and when investors behind Planet Hollywood and the Rainforest Cafe came to me to open dozens of cantinas, I sent them away without even listening to how much money I could make. The original was so special to me. But Marrandino convinced me we didn't have to cookie-cutter them, so we opened the place in Tahoe (we have since also opened in Las Vegas). The Tahoe cantina comes with this great casino showroom, and my annual Cinco de Mayo run with the Wabos in Tahoe has become a high point on my calendar every year. Tahoe has a new Sammy.

I built a brand-new studio in Marin County for the Wabos. I told them to make sure they got together and rehearsed at least once a week, but they were in and out of that studio all the time. They stayed tight. After the Van Halen reunion tour, I was so happy to get back with the Wabos. After refreshing my recollection of the pressure on the big-time rock bands, the high ticket prices, the giant production, the big crews, and all that crap, I was glad to go back to a band that can just go play. If I wanted to bring the band down to Cabo to play for free in the cantina for a week, I did. I took off basically that whole year in Cabo. I wrote my next album, *Living It Up,* pretty much about everything I was doing—"Feet in the Sand," "Living on the Coastline," all those songs.

I was still recovering from the reunion tour when Irving called to talk to me about Van Halen being inducted into the Rock and Roll Hall of Fame. At first, he said something about only the original band getting invited. I went nuts on Irving. I was in the band longer than Roth. He was in Van Halen seven years. I was with them eleven years. I sold more records than he did. How could they do this to me?

We didn't know that the brothers were fooling around with Roth again. Mikey and I were both on the outs. Irving called back and said everything was okay. It probably never was a problem. That's one of the things he does—makes problems happen so he can make them go away.

It was Van Halen, R.E.M., Patti Smith, Grandmaster Flash, and Ronnie Spector. I told Irving that we should all suck it up and make a united showing at the ceremony. He came back with their word. "If you're going, they're not going," he said. I thought they were bluffing. Right up until I was standing there giving my speech, I half-expected Roth to bust in and do something stupid. Mikey and I wanted to play. Ed and Al pulled the plug at the last minute. Velvet Revolver was set to induct Van Halen. Irving managed them, too, which may have something to do with how they landed the assignment. Since Van Halen wasn't going to perform, Velvet Revolver planned a medley of one Dave song and one Sammy song.

Roth called up Slash, the Velvet Revolver guitarist, and told him if the band played "Jump," Roth would come and sing with them. When Slash said that the band didn't have a keyboard player, Roth told him to put the part on tape. He and Slash got into it. Slash told him they were a rock-and-roll band and they played their own instruments and weren't going to pretend like they had a keyboard player just for Roth. Slash offered to play "You Really Got Me," or "Runnin' with the Devil," but for Roth, it was "Jump" or nothing. When the Velvet Revolver vocalist Scott Weiland got wind of Roth's phone call, he told Slash he would quit the band if they let that motherfucker anywhere near the stage. By the time I called Slash to suggest that Mikey and I join them for a couple of numbers, there was no way that was happening. I called Paul Shaffer, bandleader for the event, and he accepted my offer for Mike and me to do "Why Can't This Be Love" with the house band. No way I was going and not playing.

Kenny Chesney insisted on coming with me. Emeril Lagasse flew in. Mike and I were there with our wives and everybody was giving us so much love. Annie Leibovitz, the photographer, came up and hugged me. Keith Richards's daughter wanted to meet me.

"I'll take a picture with you if you take me and introduce me to your dad," I said. She dragged me over to his table.

"Hey, mate, Sammy, good job," he said. I can't tell you how I felt. It was like the first time I felt respected in this business in my life.

I did my speech. "I'm sorry the brothers and everybody's not here," I said. "God bless 'em, but you couldn't have kept me from this with a shotgun."

At the end of the evening, they built a jam around Patti Smith and her song "Power to the People." She is not my kind of gal (and I'm not her kind of guy), but I did record her song "Free Money" early in my career. Stephen Stills was up there, a little bit gone and stepping on everybody, rolling across them. Eddie Vedder was there, in great form. The R.E.M. guys were there. At the end of the jam, I was standing next to Keith Richards. He looked at me and winked.

"Good job, Sammy," he said. "Good job, mate. Congratulations." Pretty cool.

Still the Van Halens wouldn't leave me alone. That fall of 2006, I decided to go out with the Wabos and book small theaters, underplay all my best markets, no more than two shows in each city, a special treat for my hard-core fans. Irving and I discussed the strategy. He agreed that it would help stir up excitement, and the music business desperately needed some excitement. Irving thought it would help business the next time through the markets. We booked the whole tour.

It was around then we started to hear about another Van Halen reunion with Roth. We didn't believe it would happen. Mikey didn't think so, but he was out, replaced by Eddie's sixteen-year-old son, Wolfie, and, out of nowhere, the reunion was on. The label threw together another greatest-hits package, all Roth-era tracks. Irving wound up acting as the band's manager for the tour and he put Van Halen right on top of me.

They played the same cities the same week. Either they had just left town or I had just left. It was as if we were on tour together. We did great anyway, sold out all our dates, but it was such a chicken-shit move. Obviously, Eddie and Dave made him do it. The fans had waited a hundred years for the reunion with Roth and all the radio stations were talking about it. When I asked him about it, Irving acted like it was no big deal. He told me the same thing he said about the Sam and Dave tour—let them see how much better you are. When I asked him if he would send the Eagles out on top of Don Henley, he said that would be fine—it would make people talk about Henley.

As much as I respect him and as smart as he is, Irving couldn't help himself. He was making a lot more money on that big Van Halen reunion tour than off me playing those theaters. I needed a new manager. I went back with Carter, the guy who'd signed me to my first record deal. We'd never lost touch and he had been very successful in the management field since his years at Capitol—most recently, at that point, with Grammy-winning singer-songwriter Paula Cole.

WHILE ALL THESE tours were happening, the tequila business had been mostly running itself, but it had grown to the point where I could see it needed some proper management. My accountant took over the business, and he fielded an offer from this big-shot investor Gary Shansby, who had billions of dollars to spread around and already owned companies like Johnson's Wax, La Victoria Salsa, and Famous Amos Cookies. Shansby wanted to buy the tequila company for $38 million, but it was a complicated deal. He was only going to pay me half. I would maintain a 50 percent interest in the business and, after he spent three years building up the company, we would split everything after he sold out for $160 million. When I asked him what he would do differently with the company, he said, "Put some feet on the street."

My accountant wanted to sell. I didn't. It was one of the worst deals I'd ever seen. If his plan didn't work and I wanted to buy back the company, he would make me pay some heavy-duty interest on the money he advanced me for the sale. Nothing doing. Shansby hated me for not selling and started his own tequila brand. I turned around to my accountant, now running the company, and told him to do what Shansby said he was going to do. The accountant hired a marketing person. He hired six regional sales guys in the field. He hired a manager to run the salesmen. He opened an office. He spent 4 million bucks on overhead that year. Nothing really happened.

In the meantime, as my accountant, he got me into a restaurant deal in San Francisco. He found these other two investors, and the three of them went to Mexico and met everyone at my plant, making plans to start their own brand. I began to see there might be a problem. He sat down with me and told me that he and his investors wanted to buy the company from me and they would pay $22 million. I already turned down Gary Shansby's offer of $38 million. What was this guy thinking? I fired my accountant.

He did have a piece of the company. When he sued to get back my shares in his San Francisco restaurant, I made a deal for his end of Cabo Wabo and he went away. I brought in a liquor business old-timer named Steve Kauffman to run the company. He was somebody I knew, who came from Seagram and had done some work for me as a consultant. He was going through his fourth divorce, the poor bastard. He needed a job. Once he took over, the business exploded overnight.

A little more than a year after Kauffman came to work for me, Skyy Vodka approached him to buy the company. He took a lunch with an old friend from Skyy and showed him the numbers. The guy called me up from the lunch and offered me $70 million for the company. I almost fainted.

In the liquor business, we were quiet underachievers. We had four employees. I didn't spend any money marketing because we were doing fine, growing at a nice, beautiful, slow pace. The three-year average net profit was almost $7 million a year. I was happy making that much money. I didn't need any more money. I liked keeping it guerrilla, maintaining control.

I went to the Skyy offices in San Francisco, very hip company, staffed by lots of young people. I felt at home and wanted to be involved with these guys. I told them maybe I would sell them 50 percent of the company. I went back and forth for about ten days, waking up in the middle of the night thinking, "Oh, no, I can't sell this company." I finally told them I couldn't sell. By the time I paid the lawyers, taxes, bought off my partners, I told them, all I'd have left is a chunk of money that doesn't really change my life. "What amount would change your life?" they asked.

"At least $100 million," I said.

They called back the next day and said okay.

I couldn't even calculate it. What do you do with $100 million? You can't put it in the bank. It was making me more nervous than being broke did. I changed my mind and told them, once again, I couldn't sell the company. That's when Luca Garavoglia, the young, dashing chairman of Campari Group, and his sidekick with the food stains on his shirt, Stefano Saccardi, a man so relaxed and agreeable I never realized he was Luca's attorney, showed up in Cabo for my birthday bash.

Luca could be the most impressive person I've ever met in my life. He is a brilliant, subtle, and classy human being, elegant and elevated with an extraordinary command of a wide array of details. He serves on the board of directors for Ferrari. His father died suddenly when Luca was fresh out of university, and he took over the old Italian aperitif makers. He blasted the company into the modern world. He started buying brands like Skyy Vodka, and took the company public on the Italian stock market. The Campari Group became one of the top liquor companies in the world.

Luca and Stefano told me that because they ran a public company, it would be difficult for them to acquire only part-interest in the tequila, but they could figure out a way so that I could keep 20 percent of the company. I was fine. With somebody like Luca Garavoglia owning the company, my 20 percent was going to be worth more than the 100 percent had been. Luca was the clincher. In May 2007, I made the sale.

I took my whole family on vacation to Italy for six weeks, even my brother and his family. We stayed for a week at this winery the Campari people owned in Sardinia, one of the coolest places on the planet. We went from Sardinia to southern Italy, up the Amalfi coast and through Tuscany and Chianti, all the way to Lake Maggiore in northern Italy. After a week by the lake, we went to Milan to visit Luca.

He showed me the new $100 million Campari factory. Only about five people were running the whole place with these efficient new machines that wrap and seal twenty-five hundred cases of Campari in, like, two minutes. He gave me some amazing numbers—$15,000 a second or something like that—but he wasn't bragging. He was just showing me. Twenty years ago, they probably had six thousand employees. Now they have a dozen, most in the office.

"You ride with me," he said, as we left the factory. "Let's keep talking."

We got into his Maserati Quattroporte, which is not that great a car. There were two guys with him. We got in the back. The other two guys sat in the front. They looked around, started it up, and punched it. In no time, they were hauling ass 140 miles an hour down the freeway. I felt every bump and I couldn't roll down the window. We were hot. The air-conditioning was not working too well.

"I don't like this car very much," Luca said. "I like a Mercedes, but I'm on the board of directors for Ferrari and they wanted me to drive one of their cars. The Maserati is the only four-seater. The Ferrari, it's not right for me—no four doors."

We got back to the office and went through a metal detector. The two other guys pulled out the guns they were wearing in shoulder holsters. Nobody blinked an eye. Luca's car rode so badly because it was carrying sixteen hundred kilos of bulletproofing.

A couple of years later, I started reading about the new Ferrari 599 Fiorano. Ferrari enthusiasts were comparing this to the greatest of the Ferraris—the 275GTB, the first Testarossas. I hadn't bought a new Ferrari in a long time. I decided to buy one. I went to the Ferrari dealer and the salesman took my order. When we were finished, he tells me it will be a two-and-a-half-year wait and a $300,000 premium above the sticker price. I called Luca.

"It's so funny you called me," he said. "There's a board meeting on Monday. I'll see what I can do. What exactly do you want?"

That Monday night, I received an e-mail from Luca with a letter from the CEO of Ferrari attached, for the dealership. I got my car in two and a half months and I paid sticker price. Ferrari even gave me a custom paint job, black with a red stripe. They reversed the color on the interior—red interior with black stitching. They put in a plaque that said THIS FIORANO MADE FOR SAMMY HAGAR, and delivered it on my birthday. On every other Ferrari, the symbol is always a black horse on a yellow background. They made mine red.

Life was good for a kid from the orange groves. I had wealth and fame. I was in the fucking Rock and Roll Hall of Fame. I am blessed with a wonderful wife, great children, and a loving family. I had experienced success in every realm of my life. It had been a long road to Cabo.

I went back to Fontana in February 2007 to play "I Can't Drive 55" at the NASCAR track they built on the site of the old Kaiser Steel plant. The stage stood where my father used to work at the open hearth. Family was everywhere. I got a huge ovation, bigger than the Hollywood celebrities there, whose names I couldn't remember the next day. I took Kari and the girls and we spent three days with my mother at her place in Palm Desert.

After Mike died, she sold the farm I bought them and moved to a condo in Palm Desert near my sister Velma. My mom would have rather found a five-dollar bill walking down the street, or pick up tin cans and cash them in for twenty bucks, than have me give her a million. She wanted to win the lottery or a jackpot. She didn't want anybody to give her anything.

I finally talked her into letting me buy her a house. She loved her condo. We furnished it and remodeled it, but this really nice house came up across the street from Velma. It belonged to one of the executives for Outback Steakhouse, who had been transferred to Florida. She loved it. We held big events there every year. She would cook giant dinners for Easter or Thanksgiving and we would surround her with family.

There was an extra bedroom for me and Kari. We kept some of our stuff there. I had a guitar there. She came to stay with us, too. She had her first heart attack at four in the morning while she was visiting. She finally confessed that she couldn't breathe, and we took her to the hospital. They put two stents in her, but she had another heart attack a year later. She went under the knife for quadruple bypass surgery at age eighty. I was scared. I thought she would die. But she volunteered for the operation and she came out of it great.

When we stayed with her before the NASCAR event, she seemed tired. My mom was always a ball of energy, but she was sitting a lot. She stayed up late the night before we left, talking with me. My mom always kept a garden. She saved seeds and was careful to rotate her crops in their little plots. She knew all this stuff. "You can't just take carrot seeds and plant them the next year," she said. "It has to be the third year." She talked about how peach seeds had to be wrapped with cotton and left in a jar with water all winter before they would sprout. She told me you have to plant carrots in a spot one year, and the next year put lettuce in that spot, and then the next year you leave it alone, and then you come back with tomatoes. She told me all these wonderful things about canning. I got up the next morning, went and did the NASCAR gig for my father, and went home. Two days later, she died.

My sister called at about six in the morning. She was putting Mom in an ambulance. Twenty minutes later, the phone rang again. She was dead. She didn't have a heart attack. Her heart simply stopped beating. She was done and she just slipped away quietly. "A beautiful finish," my sister Bobbi wrote. I miss her every day.

⁑16⁑

WHO WANTS TO BE A BILLIONAIRE?

I have a real hard time giving up on the record business. That has nothing to do with my livelihood. It's just such a shame. People don't realize how dried up it's become. The only people that sell records anymore are brand-new little pop bands that kids buy. For me, there is no record business. That breaks my heart. I want to make records. That's a big part of what I do. But even before Napster ruined everything, I was looking for ways around the major labels. I gave an album to the guys at Tower Records, who were starting a little label. The big companies had screwed me out of money for years. I just had been with MCA and the Bubble Factory with *Marching to Mars* and *Red Voodoo*, each of which sold nearly a half million. It was not the kind of numbers I was used to selling, but I just wanted to make records. I saw all that going in the tank. I saw record companies changing, not putting any effort into a guy like me.

All you need is one new song. I made a string of recordings, and released them as singles, like "Sam I Am" or "I'll Take You

There," the old Staples Singers song. Later I might put them on a record, but they were really little more than tiny treats for radio and my fans. I started spending my own money. I'd write a song, record it in my studio, print it up, hire a promotion man, have him ship it to all the radio stations, pay a little bit of money here and there, and get it played so that I could have a new song. That was fun, not profitable, but I make money other ways. I can have fun with my music now.

That's been the great gift of Cabo Wabo. The tequila business pays for me to still be in the music business. I could make money on tour, but I wouldn't be traveling in a private jet and staying in nice hotels when I'm an opening act or playing small venues. I wouldn't be making any records.

My band gets paid like a big-time band. I don't take much money from my musical career, which makes me love it more. It takes the business out of it. That's been the most successful thing I've done, take the business out of my music, because now, anything I do comes from my heart, baby. I am also free to do things like the Staples Singers song. I love to not be bound by my image or what people think of me. I loved that song. I was so happy to record it. It was like "Sam I Am." I felt like I could write and record anything I want. There's no industry out there anyway.

Bottom line is, I want to make records. I want to write songs. I want to go out and play new material for people. That was one of my main reasons for putting together Chickenfoot with Chad Smith on drums, Joe Satriani on guitar, and Michael Anthony on bass.

It had begun several years earlier, when Chad Smith from the Red Hot Chili Peppers started coming down to the birthday bash every year. He bought a house down there. We would jam, jam, jam. I told him to never go to Cabo without calling me first. I'd call him and say I was going to Cabo the next day and he would get on a plane and come down. He and Michael Anthony and I would play every night. We called our little trio Chickenfoot. We played cover tunes. We played any damn thing we wanted. I'd let Mike sing lead. We used to do a great medley of "Come Together" and "Give It Away" by the Chili Peppers. We did James Brown medleys, "Hot Pants," "Give It Up," "Turn It Loose," "Superbad" into "Cold Sweat." We were rocking, having a really good time. Chad kept telling us. "This is it," he said. "Let's start a band." It took me about five years before the light went on.

I promised myself I would never play with another genius guitar player, but I realized, for this band to work, it needed some kind of superstar guitarist. I can't play guitar and sing. That is way too much work. I can jam, but if I had to play everything every night, I'd be struggling a lot of the time. The first person I thought of was Joe Satriani.

I had called Satriani about playing together a number of years before, and he told me, "I don't play other people's music." But I'd seen him a few times since and he seemed pretty cool, quiet and aloof, distant, not too outgoing. This time, he was much more interested.

Super Bowl weekend 2008 in Las Vegas, and I had a gig booked at Pearl, the big room at the Palms Hotel. Serious fun. We were sold out. Joe came home from tour. We picked up his amp and went to Vegas. Mike and Chad met us there. We talked it over in the dressing room—just talked though. We didn't rehearse. We decided we all knew "Goin' Down," Traffic's "Mr. Fantasy," and Led Zeppelin's "Rock and Roll." We didn't discuss arrangements or anything. We just went out and did it. I did the show with the Wabos, and for the encore, I announced we were going to have some special guests, while the crew laid out Joe's gear and adjusted the drum kit for Chad. Mike's amp was already set up.

From the first thirty seconds of "Goin' Down," that audience went through the roof. I felt it. Everybody onstage felt it. Our wives, the crew, management—they all knew. Everybody in the hall felt it. It was electric. I had just done a great show with my band, played every hit you could possibly play, but this band did something else to that audience that was unique. This was a real band.

From that first time we played, I knew we had stumbled on an impossible combination of musicianship and chemistry. The cool thing about Chickenfoot was not just that the chemistry was instantly right, but also that we were all grown-ups, with our own careers, our own money. No one needed it. We played music that we liked. We weren't trying to be like this or be like that. We were exactly who and what we were, and we let whatever that was happen. That made it like jazz, in its own way. It wasn't jazz, but it was like jazz, in the sense that we were four guys playing exactly the music we wanted to play, the way we wanted to play it. For us, it worked.

I knew a so-called super-group would get more attention. Record companies could get behind it, and we could go out and play all new stuff, give me a break from "I Can't Drive 55." I love those songs, but I don't enjoy beating them to death. The first Chickenfoot record did well. It went gold. We made a profit. I have a hard time spending a few hundred thousand dollars on a record and only make a hundred back, but that's the way it is. The record industry has really died off.

I go solo backward. People like Phil Collins or Peter Gabriel leave a band to go solo. I've always gone solo first and then joined a band. They go solo because they are tired of being in a band. For me, I join bands because I need the inspiration. I learned so much in Montrose—how to play guitar like Ronnie, how to lead a band, how bands work—and I used it all in my solo career after Montrose. Ten years later, I was selling out multiple arenas, had five platinum albums in a row on Geffen, and I was ripe to join

Van Halen when they asked. I was tired of doing my own thing, thinking of taking a year off, and didn't know what to do anyway. I needed to be around other musicians to make me grow again. Ten years with Van Halen, and I was ready to go solo again. Too bad I stayed that eleventh year. I'd already learned everything I could from that band.

I wanted one last hurrah. I have been threatening to retire since 1984. Something always comes along that keeps me from quitting. Chickenfoot is the only new classic rock band in a long time. We fortunately avoided being called a super-group. That's always impossible to live up to. We certainly were super players, everyone in the group. When we got together, Joe would start playing a song idea. Chad would chime in. Mikey learns faster than anyone—he'd act like he knew the song already. I would scat along, make up lyrics on the spot or write words later. That's how we wrote every song. Joe would bring a new idea. By the end of the day, I had my part. Everybody knew what he was going to play. Magical performance. No teeth pulling. No having to extract anything out of anyone. With Chickenfoot, we simply went in and ripped it up.

We decided we wanted our record to sound classic. We thought about working with Brendan O'Brien or one of the other hot young producers, but we hired Andy Johns, even though I'd refused to work with him on *For Unlawful Carnal Knowledge*. As much as I wanted to kill him, and eventually fired him from the Van Halen record, he made the greatest classic rock records ever, *Led Zeppelin III, Exile on Main St.* Joe had worked with him, too. We both knew he came with a lot of personal problems, but when Andy Johns sets microphones up around a drum set, it fucking starts rocking. The Chickenfoot album was one of only twelve albums to go gold in 2009.

With Chad's whacked-out energy, I became the straight man. Anyone else would lose the groove twenty times by turning the beat around as often as he does, but he keeps the groove and plays everything you can possibly play in every song. He's one fierce drummer. He brings Joe out, makes Joe not be so conservative. And Joe can be a technician's technician. But not with Chickenfoot. There is nothing like the 'Foot. Chickenfoot is never going to make me a billionaire. Why am I doing this? Working so hard?

If I wanted to be a billionaire, I probably could. I could take all I have right now and leverage it like Donald Trump. That's what he did. I could work for it, too. Another couple good ideas and who knows? I could invent so many different places that are fun, which would be successful, that I'm sure I could do it, if I wanted. It's probably not that hard. But I'm not interested. I don't want to be a billionaire. That would be the biggest fucking waste of time on the planet.

When I came up with Sammy's Beach Bar & Grill, I knew it was a great idea, but I had to wonder why I would do it. I didn't need the money. I learned I can do it and how to do it, and it seemed like the easiest thing in the world for me to do. I just couldn't think of a single reason to do it. Then I thought I could do it for charity. Kari and I like to support charities that help children.

I'm involved in this charity with the Beach Bar & Grill in Maui, helping desperately ill kids on the island. All our Beach Bar & Grills support charities in the cities where they're based. We helped pay for a kidney transplant for this girl. We didn't pay for the surgery, but we paid for her family to come with her to San Francisco, where she was having the operation. We built a number of these airport restaurants—Las Vegas, St. Louis, JFK Airport in New York—and they kick down a steady stream of income to local charities working with children. This guy I work with on the Maui charity asked me if I would talk to these teenagers that had been in orphanages. They get released and they don't have anybody. They don't have a brother or a sister. They don't know where their mom and dad are. He said these kids get in trouble. They don't have any other choice. He runs a shelter for them. They don't have any self-esteem, he says. They don't believe in themselves one bit. I can fix that. I can help these kids, because I can say to them, first of all, no matter who your parents are, no matter where you're born, everyone is born with the same gifts. You've got the same ability as I have, as anybody has. Just because you don't know who your parents are, don't think that you don't have the ability to get whatever you want.

You are just as powerful as me or anybody else. You're born with that. I know it's different for you in your heart, but I'm telling you, you were born with the same gift I was, the same gifts as the Kennedys or anybody else. You can get a cold, like everybody else. You can get a job, like anyone else. You can have a family. You can fall in love. You can fall off a cliff. You can be like everyone else. Nothing's different for you. You can be as much a success as anybody. And I know there is no end to success.

I definitely believe in God and, even if I didn't, I believe that you should. People have to believe in something. Without belief in something, you're just going in circles. I do believe in spiritual things. I believe in the unknown. I believe in God and I believe in UFOs and aliens and all that mystery. I'm a big sucker for all of that. One of my main issues is kindness and being a good person in life. I don't believe in killing people, inflicting your will on another person and trying to hurt them in any way. I don't care how bad they are—it's not your business and you fuck around with that, you're fucking around with evil. If people knew how sacred life is and wouldn't take another person's life, we'd be a much fucking better-off race. This planet would be in better shape.

I won't do certain things. I know my fans see something in me that makes them respect me. I feel like I'm not worthy sometimes. I know I'm a good guy, that I mean well and give a lot of myself, but they see something in there, I guess, that's deeper than what I think of myself. I think of myself as a good guy, but I don't see myself as special. Unique, maybe, but I don't see some big star when I look in the mirror in the morning. I see my mom. She was the salt of the earth. She'd dig through Dumpsters. When my mom used to take us out on weekends, her idea of going and doing something as a family was to take us four kids to the city dump and rummage. And we loved it. It was a blast, finding stuff. It was like a treasure hunt to us kids. My mom would find food, like rotting fruit and vegetables that grocery stores were dumping. She would find two or three good oranges and be so happy. "Oh, look at these!" she would say.

I bought her houses and cars. There was nothing I wouldn't do for her, but she wouldn't have it. I had to practically force her to let me buy her anything. "Mom," I'd say, "just tell me—what do you want? What's something you always wanted to do?"

"Well, I don't know," she'd say. "I have pretty much everything I could want."

She was so solid. I'm like that. I really don't feel like somebody. I don't feel like some big star and I don't want to be some billionaire. I have all these crazy ambitions, but there's something inside of me that is my mom, and I really like that.

NOTE FROM THE COAUTHOR

I feel lucky and privileged to help Sammy tell his remarkable story. Sam and I go way back—I saw the Justice Brothers at the Wharf Rat—and he once gave out my home phone number to a sold-out Cow Palace audience, taking exception with one of my reviews. As a lifelong card-carrying member of rock music's critical elite, with its carefully proscribed doctrinaire orthodoxy, I am fully aware of the regard in which such circles hold Sam. To them, I say, fuck you, the guy had "Rock Candy" on his first album. There are entire highly regarded careers that have never reached such a peak, and that was just Sammy's opening salvo. Those who have known Sammy all these years will further testify that he has unwaveringly been the same guy all along. He is as authentic as the Grateful Dead, maybe more. Sammy Hagar is a genuine working-class hero. I am as proud of this book as anything I have ever done.

—J.S.

ACKNOWLEDGMENTS

This book is for my mom.

Thanks to my wife and children—Kari, Aaron, Andrew, Kama, and Samantha; my brother, Bob Jr. (Ponchito); my two sisters, Bobbi and Velma; Betsy and Bucky; all my aunts and uncles, nieces, nephews, cousins, and in-laws who help make my life so colorful; my band the Wabos—Vic, David (Bro), Mona, and Mikey; Renata and Bill Ravina; Carter; Joel Selvin for talking me into doing this book and finally coming over to my side; Ronnie Montrose; Ed and Al; Joe and Chad; my crew—Paul, Rosie, Ace, Jim, Three, Big Kenny, Chris, Rick, Gage, Duggie, Manning, Rich, and Austin; all the employees from the Cabo Wabo Cantinas and Sammy's Beach Bar & Grills; Marco and Jorge; Dick Richmond for helping write this book; Don Marrandino; Stan Novak; Frank Sickelsmith; Don Pruitt; John Koladner; Gary Arnold; Ed Leffler, my second father; Shep Gordon; Steve Kauffman; the Skyy and Campari team; Wilson Daniels. All my chef buddies, all the musicians that have played the Cabo Wabo, and the musicians I've had the pleasure of playing with. Ma, Lu, and Chick; all of my old buddies I grew up with; stepfather Mike for buying me my first car; and a special thanks to the fans, all the Red Heads, for being the best travel companions any artist can have. And my father, who believed in me more than himself.

—SAMMY

I NEED TO thank Lisa Sharkey, Matt Harper, and all the Harper-Collins/It Books team; Frank Weimann of the Literary Group (and Elyse Tanzillo); Carol Mastick for the transcriptions; Peter Riegert and the entire staff of the Pierre for the hospitality; Carter for saying no; David Ritz for showing me how; and, always, Carla for being herself. And, of course, my biggest thanks to Sammy.

—J.S.